AWS CDK in Practice

Unleash the power of ordinary coding and streamline complex cloud applications on AWS

Mark Avdi

Leo Lam

BIRMINGHAM—MUMBAI

AWS CDK in Practice

Group Product Manager: Preet Ahuja

Publishing Product Manager: Niranjan Naikwadi

Content Development Editor: Sujata Tripathi

Technical Editor: Rajat Sharma

Copy Editor: Safis Editing

Project Coordinator: Ashwin Dinesh Kharwa

Proofreader: Safis Editing

Indexer: Tejal Daruwale Soni

Production Designer: Shankar Kalbhor

Marketing Coordinator: Rohan Dobhal

First published: June 2023

Production reference: 1180523

Published by Packt Publishing Ltd.
Livery Place
35 Livery Street
Birmingham
B3 2PB, UK.

ISBN 978-1-80181-239-9

www.packtpub.com

To my beloved daughter Olivia, you are the light of my life and the source of my inspiration. Your unwavering love and boundless curiosity inspire me to be a better person every day. To my parents, Nasrin and Fahim, I am forever grateful for your unwavering belief in me and for instilling in me the values of hard work, determination, and perseverance. This book is a tribute to your unwavering love and guidance, and I dedicate it to you with all my heart.

– Mark Avdi

Thank you to each and every one of the hardworking individuals at Westpoint, who have fully invested in our motto, "Build, deliver, and grow."

And a very special thanks to Rico, Logan, and Roman; you give me the daily drive to be the best version of myself I can be. Thank you.

– Leo Lam

Contributors

About the authors

Mark Avdi is a seasoned technologist with a passion for building scalable cloud solutions and improving software productivity. In addition, he has extensive experience in software engineering, having worked in various industries such as finance, automotive, and education. Throughout his career, Mark has led and delivered numerous complex software solutions. His expertise and knowledge have been invaluable to the success of the Westpoint team and the technical direction of the company. As the **chief technical officer** (**CTO**) of Westpoint Software Solutions, Mark is responsible for overseeing the technical aspects of the company and guiding the development of cutting-edge software solutions.

We would like to extend our appreciation to all members of the Westpoint team whose dedication and hard work continue to inspire us to push the boundaries of what's possible. Special thanks go to **Hugo Bonatto** *and* **Gabriel Correia** *for their contributions to the book, which were invaluable in making it a success.*

Leo Lam is a rugby player, coach, and team builder. A solutions architect by trade, Leo has utilized his expertise in the world of complex software architecture to that of an architect of technology teams. Drawing from the disciplined, hardworking, and delivery-focused ethos of the rugby teams he has played in and coached, Leo has implemented such values in the dedicated group of engineers at Westpoint where he is the team operations lead.

Leo has extensive experience within the world of cloud development and architecture, specifically AWS and AWS **Cloud Development Kit** (**CDK**). He has been on a mission to infuse the already knowledgeable team of engineers at Westpoint with efficient and modernized ways of development using infrastructure as code. This has allowed the Westpoint team to efficiently and effectively deliver highly complex and large-scale enterprise solutions at warp speed.

Nothing means more to Leo than a team that has discipline, accountability, and focus to deliver exactly what needs to be delivered.

About the reviewers

Dhiraj Khodade is an experienced, hands-on software architect and cloud engineer involved in the end-to-end architecture, design, and implementation of cloud-native applications. He has received a bachelor's degree in computer science from the University of Pune in India. He has been in the IT space for 10 years.

Having started working on AWS CDK close to its inception, he loves the concept of **Infrastructure as Code (IaC)** using object-oriented programming and has used it to provision and maintain enterprise cloud platforms.

Dhiraj is a technophile who focuses on cutting-edge technologies and new platforms to stay at the forefront of the IT industry. Dhiraj likes to travel. He sometimes writes technology blogs. He lives with his wife in Tokyo, Japan.

I would like to thank my parents and my wife, Anjali, for their support. And special thanks to the AWS CDK open source community and experts for constantly putting their best foot forward each day, which has helped me avoid the tangled mess of CloudFormation and its maintenance nightmare.

Igor Soroka is the co-founder of and a consultant at Soroka Tech. He has spent the last three years working with serverless projects on AWS as a hands-on consultant, technical lead, and coach. Igor is a versatile cloud professional helping teams and companies of various sizes to change their ways of developing and deploying web applications. He prioritizes cost optimization, security, and high availability in his work. Igor has worked with start-ups, scale-ups, and Fortune 500 companies. Also, he shares his knowledge through blogs, talks, and podcasts as an AWS Community Builder.

Thank you to my family, who have been a pillar of strength, encouraging me every step of the way and providing me with the love and support that has helped me to achieve my goals. I also want to thank my cat, who has been a constant source of joy and comfort. His affectionate purrs and playful antics have made me smile even on the most challenging days, and I am grateful for his companionship.

Mischa Spiegelmock is a technical co-founder of multiple successful software start-ups in San Francisco, with experience in the tech industry since beginning his career at LiveJournal in 2005. He co-founded DoctorBase in 2009, a healthcare technology start-up, and JetBridge in 2017, a software consultancy focusing on building cloud-native applications and connecting top engineers worldwide with Silicon Valley start-ups. Mischa is a regular open source contributor and the author of the *Leap Motion Essentials* book from Packt, and has written and spoken on various technical topics to educate others.

Table of Contents

Part 2: Practical Cloud Development with AWS CDK

3

4

5

6

Testing and Troubleshooting AWS CDK Applications 87

Part 3: Serverless Development with AWS CDK

7

Serverless Application Development with AWS CDK 109

8

Streamlined Serverless Development 129

Part 4: Advanced Architectural Concepts

9

Indestructible Serverless Application Architecture (ISAA)   145

10

The Current CDK Landscape and Outlook   155

Preface

Welcome to our book on the exciting world of **Infrastructure as Code (IaC)** using **AWS Cloud Development Kit (CDK)**. In this book, we will take you on a journey through the world of modern cloud infrastructure and show you how to leverage the power of AWS CDK to create and manage your infrastructure using familiar programming languages.

AWS CDK has revolutionized the way developers think about IaC by providing a more flexible, scalable, and automated approach to infrastructure management. With AWS CDK, you can programmatically define and manage your cloud infrastructure, making it easier to adopt DevOps practices and scale your applications.

Throughout this book, we will cover the core concepts of AWS CDK and show you how to use it to build and deploy various aspects of cloud applications. We will also cover advanced topics, such as testing, troubleshooting, and best practices for managing large-scale infrastructure.

Who this book is for

Whether you're a solutions architect, a seasoned developer, or just starting your journey in the cloud, this book is for you. With our practical approach and easy-to-follow examples, we'll help you unlock the full potential of AWS CDK and take your infrastructure to the next level.

What this book covers

Chapter 1, *Getting Started with IaC and AWS CDK*, will cover setting up your local machine for CDK development, introduce you to AWS CDK, and dive into its inner workings.

Chapter 2, *A Starter Project and Core Concepts*, will cover CDK integration using the monorepo model and core CDK concepts such as constructs, creating custom constructs, and community-maintained and construct libraries.

Chapter 3, *Building a Full Stack Application with CDK*, will guide you through building a full stack application using AWS CDK, Node.js, Express.js, React, ECS, and DynamoDB. You'll learn how to create a backend API with Express.js, connect it to a React frontend, and deploy it using AWS CDK. Additionally, you'll see how AWS CDK simplifies the process of building and deploying Docker images to AWS ECS.

Chapter 4, *Complete Web Service Deployment with AWS CDK*, will cover how to set up DNS for frontend and backend URLs with Route 53, create an AWS RDS MySQL database using AWS CDK and secure endpoints with AWS ACM TLS certificates, and set up CloudFront distribution for frontend assets.

Chapter 5, Continuous Delivery with CDK-Powered Apps, will provide an introduction to CI/CD and AWS's toolset, creating different environments for our application, and using AWS's CodeBuild and CodePipeline to implement a robust CI/CD process. We will also go over running the build for various branches and getting notifications of the build status.

Chapter 6, Testing and Troubleshooting AWS CDK Applications, will cover automated testing for AWS CDK applications. We'll cover different types of tests, strategies for writing effective tests, and best practices for integrating testing into your AWS CDK development workflow.

Chapter 7, Serverless Application Development with AWS CDK, will cover creating a serverless application using AWS CDK. Topics include setting up an API Gateway, Lambda functions, a Step Functions state machine, and a DynamoDB table.

Chapter 8, Streamlined Serverless Development, will cover common serverless development issues and explore solutions for running Lambda application logic and AWS services locally. We'll use tools such as LocalStack to simulate AWS services and integrate our Lambda functions with a local express server.

Chapter 9, Indestructible Serverless Application Architecture (ISAA), will talk about the **Indestructible Serverless Application Architecture** (**ISAA**), which uses AWS CDK to create highly resilient, scalable, and maintainable cloud applications. The principles of ISAA include a fuller stack, a serverless architecture, simplicity, a single-table design, and an event-driven architecture. This chapter is mostly theoretical and covers the principles of ISAA, technical requirements, and example scenarios.

Chapter 10, The Current CDK Landscape and Outlook, will give you an idea of the current landscape and outlook for AWS CDK, including how it has revolutionized the IaC landscape and enabled dynamic provisioning. This chapter also covers the use of AWS CDK in large organizations and explores CDK alternatives, such as Pulumi and CDKTF, as well as CDK-inspired projects such as CDK8S.

To get the most out of this book

You should have a basic understanding of cloud computing concepts and the services provided by major cloud providers, such as AWS. Familiarity with at least one programming language, preferably TypeScript, and some experience with writing scripts would also be helpful. Additionally, you should have an interest in IaC and be familiar with its basic concepts. Finally, some understanding of DevOps practices would be beneficial.

Technologies covered in this book
TypeScript 3.7
AWS CDK
Docker
Bash
Operating system requirements
Windows, macOS, or Linux

If you are using the digital version of this book, we advise you to type the code yourself or access the code from the book's GitHub repository (a link is available in the next section). Doing so will help you avoid any potential errors related to the copying and pasting of code.

Download the example code files

You can download the example code files for this book from GitHub at `https://github.com/PacktPublishing/AWS-CDK-in-Practice`. If there's an update to the code, it will be updated in the GitHub repository.

We also have other code bundles from our rich catalog of books and videos available at `https://github.com/PacktPublishing/`. Check them out!

Code in Action

The Code in Action videos for this book can be viewed at `https://packt.link/sXFWF`.

Download the color images

We also provide a PDF file that has color images of the screenshots and diagrams used in this book. You can download it here: `https://packt.link/EQ3qe`.

Conventions used

There are a number of text conventions used throughout this book.

`Code in text`: Indicates code words in text, database table names, folder names, filenames, file extensions, pathnames, dummy URLs, user input, and Twitter handles. Here is an example: "If you dig into any of these directories, you will see that they each has its own README files, `package.json` files, and various other relevant configurations for building high-level components."

A block of code is set as follows:

```
useEffect(() => {
    const fetchTodos = async () => {
      const response = await axios.get(backend_url);

      setTodos(response.data.todos);
    };

    fetchTodos();
}, []);
```

Any command-line input or output is written as follows:

```
$ cdk deploy --profile cdk
```

Bold: Indicates a new term, an important word, or words that you see onscreen. For instance, words in menus or dialog boxes appear in **bold**. Here is an example: "Click on **Request**, select **Request a public certificate**, and click **Next**."

> **Tips or important notes**
> Appear like this.

Get in touch

Feedback from our readers is always welcome.

General feedback: If you have questions about any aspect of this book, email us at customercare@packtpub.com and mention the book title in the subject of your message.

Errata: Although we have taken every care to ensure the accuracy of our content, mistakes do happen. If you have found a mistake in this book, we would be grateful if you would report this to us. Please visit www.packtpub.com/support/errata and fill in the form.

Piracy: If you come across any illegal copies of our works in any form on the internet, we would be grateful if you would provide us with the location address or website name. Please contact us at copyright@packt.com with a link to the material.

If you are interested in becoming an author: If there is a topic that you have expertise in and you are interested in either writing or contributing to a book, please visit authors.packtpub.com.

Share Your Thoughts

Once you've read *AWS CDK in Practice*, we'd love to hear your thoughts! Scan the QR code below to go straight to the Amazon review page for this book and share your feedback.

https://packt.link/r/180181239X

Your review is important to us and the tech community and will help us make sure we're delivering excellent quality content.

Download a free PDF copy of this book

Thanks for purchasing this book!

Do you like to read on the go but are unable to carry your print books everywhere? Is your eBook purchase not compatible with the device of your choice?

Don't worry, now with every Packt book you get a DRM-free PDF version of that book at no cost.

Read anywhere, any place, on any device. Search, copy, and paste code from your favorite technical books directly into your application.

The perks don't stop there, you can get exclusive access to discounts, newsletters, and great free content in your inbox daily

Follow these simple steps to get the benefits:

1. Scan the QR code or visit the link below

https://packt.link/free-ebook/978-1-80181-239-9

2. Submit your proof of purchase

3. That's it! We'll send your free PDF and other benefits to your email directly

Part 1: An Introduction to AWS CDK

This part introduces AWS **Cloud Development Kit** (**CDK**) as a cloud provisioning tool. We will immediately get into the action by provisioning a container service on AWS using CDK, and then we will cover some of the core concepts before delving further into more advanced topics. This part has the following chapters:

- *Chapter 1, Getting Started with IaC and AWS CDK*

- *Chapter 2, A Starter Project and Core Concepts*

1

Getting Started with IaC and AWS CDK

Infrastructure as code (IaC) is the standard practice of defining various infrastructure components using code. You must have heard about a few of these software tools, such as Chef, Puppet, Terraform, or the old-school Bash programming, to set up servers in a predictable way.

The need for IaC was born out of the toolset for spinning up servers circa 2006 lacking the predictability, speed, and agility of the code used in programs that were deployed onto servers. Pre IaC, the normal workflow of deploying something on AWS would be this:

1. Spinning up **Elastic Compute Cloud** (EC2) servers manually via the dashboard

2. SSHing into the machines

3. Copying some Bash configuration files using **Secure Copy Protocol** (SCP)

4. Running them in a certain sequence to deploy an application

5. Configuring database connections

This is one of the better scenarios of how things were done back then.

Needless to say, this method was not scalable, repeatable, or reliable enough to deal with the ever-increasing complexity of web applications on the cloud. Around the same time, tools such as Chef, Puppet, and a few others showed up, which helped immensely with server configuration.

A couple of years later, tools such as Terraform popped up, which at the time were quite revolutionary in the way that they defined the desired state of the deployment on a given **cloud service provider** (CSP) such as **Amazon Web Services** (AWS) in a declarative fashion. This configuration would then be fed into the Terraform toolset, which would then decide which AWS services needed to be created, deleted, or modified. Let's look at one such Terraform configuration file:

```
resource "aws_instance" "single-instance"{
   ami             = "ami-ebd02392"
   instance_type = "t2.micro"
```

```
    subnet_id       = "subnet-eddcdzz4"
    vpc_security_group_ids = ["sg-12345678"]
}
```

The preceding Terraform configuration file defines an AWS EC2 server using the `aws_instance` resource type provided by Terraform's AWS provider and then is given an **Amazon Machine Image** (**AMI**) configuration (which is essentially the operating system image) instance type to define its CPU and RAM configuration and other network settings. If you have Terraform set up correctly locally, this will investigate your AWS setup, see that this server doesn't exist, and then proceed to create this server for you.

Obviously, this is a very simple example. There are Terraform configuration projects with many thousand lines of configuration files. These projects can get quite elaborate and out of hand once the complexity of the deployment increases. There are various remedies for this, none of which—in our opinion—really addresses the issue here. Have you spotted the problem?

The problem is that these configurations are not really defined as code. They are defined as declarations in a declarative language that are not Turing Complete code blocks.

In this chapter, we will cover the following main topics:

- Introduction to AWS CDK
- Setting up your local environment and writing your first CDK app
- Creating a containerized web application in AWS CDK using Docker
- Understanding the inner workings of AWS CDK

Technical requirements

You will be able to find the code associated with this book in this GitHub repository: `https://github.com/PacktPublishing/AWS-CDK-in-Practice`.

This repository is organized into directories for each chapter. This code in this chapter can be found in this directory: `https://github.com/PacktPublishing/AWS-CDK-in-Practice/tree/main/chapter-1-getting-started-with-iac-and-aws-cdk`.

The code examples provided in this book are written in TypeScript. **Cloud Development Kit** (**CDK**) supports JavaScript, Python, Java, and C#. We're sticking with TypeScript since it's the best supported of the bunch by the CDK development team, and every feature, doc, or material is first released on TypeScript and then translated into other programming languages.

That said, converting the TypeScript code samples in this book into the programming language of your choice is relatively easy since the framework follows a set of common interface definitions. AWS has a handy guide for you: `https://docs.aws.amazon.com/cdk/v2/guide/multiple_languages.html`.

Do also note that to work with CDK, you will need an AWS account. You can sign up for one here: `https://portal.aws.amazon.com/billing/signup`.

For more advanced users, we recommend setting up a different AWS organization for working with CDK while you're learning the ropes so that your development doesn't interfere with applications that you might have already spun up in your account. You can find out more about AWS Organizations here: `https://aws.amazon.com/organizations/`.

We also highly recommend setting AWS billing alarms. We will keep things within the free tier of various services as much as we can in this book and will destroy stacks as soon as we've spun them up and interacted with them. But it's best to be on the safe side and set a spending alarm in case mistakes happen or you forget to destroy stacks. A spending alarm of $10 should do the trick. The following link describes how you can do this: `https://docs.aws.amazon.com/AmazonCloudWatch/latest/monitoring/monitor_estimated_charges_with_cloudwatch.html`.

The Code in Action video for this chapter can be viewed at: `https://packt.link/nX11h`.

Introduction to AWS CDK

In the introduction to this chapter, we covered a basic history of IaC and how these software tools lack the expressiveness of modern programming languages. AWS without a doubt is the frontrunner in the cloud industry, and its answer to difficult infrastructure setup is CDK.

In AWS's own definition, CDK is an open source software development framework—notice how AWS is not calling it an infrastructure provisioning framework—that defines AWS resources for a cloud application using familiar programming languages.

AWS's definitiomeans, for example, that instead of defining the EC2 instance using declarative configuration files as we did with Terraform's **HashiCorp Configuration Language** (HCL) in the intro, you can define the same instance as the following:

```
import * as cdk from '@aws-cdk/core'
import * as ec2 from '@aws-cdk/aws-ec2'

 export class MyAppStack extends Stack {
  public readonly default_vpc: IVpc;

  public readonly my_virtual_machine: Instance;

  constructor(scope: Construct, id: string, props?: StackProps) {
    super(scope, id, props);

    this.default_vpc = Vpc.fromLookup(this, 'VPC', {
      // This will get the default account VPC
      isDefault: true,
    });
```

```
    this.my_virtual_machine = new Instance(this, 'single-intance', {
      // The type of instance to deploy (e.g. a 't2.micro')
      instanceType: new InstanceType('t2.micro'),
      // The type of image to use for the instance
      machineImage: new AmazonLinuxImage(),
      // A reference to the object representing the VPC
      vpc: this.default_vpc,
    });
  }
}
```

> **Note**
> Don't worry about running the code in this section; we will get into that after we've completed setting up your local machine for CDK development.

You might (rightfully) think that in terms of lines of code, this doesn't really save you much time. It's more or less the same amount of configuration we declare. Let's see what happens if we want to declare 100 EC2 instances:

```
import * as cdk from '@aws-cdk/core'
import * as ec2 from '@aws-cdk/aws-ec2'

export class MyAppStack extends Stack {
  public readonly default_vpc: IVpc;

  public readonly my_virtual_machines: Instance[];

  constructor(scope: Construct, id: string, props?: StackProps) {
    super(scope, id, props);

    this.default_vpc = Vpc.fromLookup(this, 'VPC', {
      isDefault: true,
    });

    this.my_virtual_machines = [...Array(100).keys()].map(
      i =>
        new Instance(this, `single-intance-${i}`, {
          instanceType: new InstanceType('t2.micro'),
          machineImage: new AmazonLinuxImage(),
          vpc: this.default_vpc,
        }),
    );
  }
}
```

Are you excited yet? Let's now imagine we want to deploy 100 EC2 instances and we want every odd one to be a t2.large instance type instead. Let's also throw in a **Simple Storage Service (S3)** bucket too and give all the programs that run within our VMs read access to some files in this S3 bucket. Perhaps there is a shell script in there that we want them to run as soon as they spin up. The following CDK code does exactly that:

```
import { RemovalPolicy, Stack, StackProps } from 'aws-cdk-lib/core';
import { Instance, InstanceType, AmazonLinuxImage, Vpc, IVpc } from
'aws-cdk-lib/aws-ec2';
import { Bucket } from 'aws-cdk-lib/aws-s3';
import { Construct } from 'constructs';

export class MyAppStack extends Stack {
  public readonly bucket: Bucket;

  public readonly default_vpc: IVpc;

  public readonly my_virtual_machines: Instance[];

  constructor(scope: Construct, id: string, props?: StackProps) {
    super(scope, id, props);

    this.default_vpc = Vpc.fromLookup(this, 'VPC', {
      isDefault: true,
    });

    this.my_virtual_machines = [...Array(100).keys()].map(
      i =>
        new Instance(this, `single-intance-${i}`, {
          instanceType: new InstanceType(i % 2 ? 't2.large' : 't2.
micro'),
          machineImage: new AmazonLinuxImage(),
          vpc: this.default_vpc,
        }),
    );

    this.bucket = new Bucket(this, 'MyBucket', {
      removalPolicy: RemovalPolicy.DESTROY,
    });

    this.my_virtual_machines.forEach(virtual_machine => {
      this.bucket.grantRead(virtual_machine);
    });
  }
}
```

As you can see, CDK is an extremely versatile IaC framework. It unleashes the power of programming languages and compilers to declare highly complex AWS infrastructure with code that is, compared to the alternatives, easily readable and extensible. We can loop, map, reference, write conditions, and—more importantly—use the helper functions CDK provides to simplify the previously daunting task of implementing IaC.

This is a revolutionary concept. Full stack developers no longer must learn convoluted **domain-specific languages** (**DSLs**) to interact with AWS services; they can bring their current development skills to the world of cloud provisioning. There is no longer a context switch between development and DevOps. You no longer need to leave your IDE, and you can build and deploy the CDK infrastructure using common tools you've previously used—for example, npm.

Additionally, CDK has powerful customization and component sharing baked in. You can create batches of stacks, with your own set of governance and compliance requirements as reusable components. A set of certain EC2 machines with AWS Systems Manager parameter store configuration, security groups, certificates, and load balancers that is common within your stack can then be defined as a component (or in CDK terms, a construct) and then initialized and reused as many times as you desire. This is the true power of CDK, and it unleashes countless new opportunities for developers, as we will explore in the following chapters.

Till now, we've learned why CDK is cool. We hope you grow to love the toolset just as much as I do as you apply it to your own programming stack. Let's now get into the good bits and start off by setting up your local machine for CDK development.

Setting up your local environment and writing your first CDK app

AWS CDK comes with two important tools, as follows:

- The AWS CDK v2 standard library, which includes constructs for each of the AWS services
- The CDK CLI, which is a command-line tool for interacting with apps and stacks

Before setting these up, let's go ahead and set up AWS's main CLI application to aid us with our CDK setup and deployments later on.

Setting up the AWS CLI and profile

To use CDK to provision AWS services, we will first need to set up our local machine with the configuration it needs to work with AWS. The easiest way to do this is via the AWS CLI.

> **Note**
> The AWS CLI is different from the AWS CDK CLI. The AWS CLI is mainly used to interact with AWS via command-line or Bash scripts, but it also comes in handy when using CDK.

AWS has detailed AWS CLI installation documentation for various operating systems and processor architectures, which can be found by visiting the following link: `https://docs.aws.amazon.com/cli/latest/userguide/getting-started-install.html`.

Follow the steps required for your operating system and come back here once you're done.

We hope that wasn't too painful. we're on macOS, which seems to have the fewest steps of them all. Next, let's check the installation was successful by running the following command:

```
$ aws --version
```

If all has gone well during the installation process, you should be seeing an output like the following screenshot:

Figure 1.1 – AWS CLI version output

Next, let's create a separate profile for this book. As mentioned in the *Technical requirements* section, We highly recommend you create a separate account while you're developing with CDK rather than using the account you normally use in your day-to-day activities.

Log in to the AWS account of your choosing, and once logged in, type IAM in the search box at the top of the AWS dashboard:

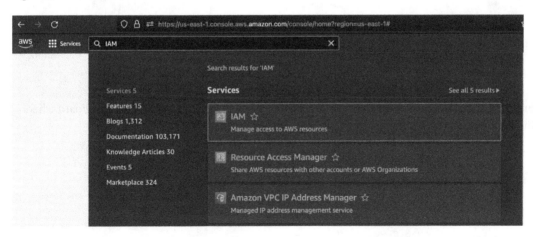

Figure 1.2 – Finding IAM in the sea of AWS services

Click on **IAM** next, and from the menu on the left, select **Users**:

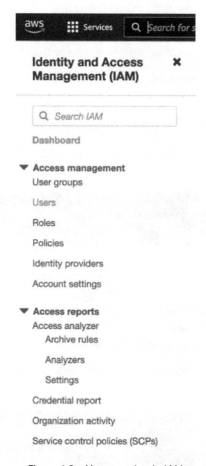

Figure 1.3 – Users section in IAM

From the list of users, click on your own username. You will be taken to the user dashboard where the **Permissions** tab is active:

Figure 1.4 – IAM user roles

As you can see in the preceding screenshot, we've given our user **AdministratorAccess** permissions so that we have free rein on spinning up various services via CDK. Otherwise, we will have to constantly come back here and give ourselves service-specific permissions to spin up services.

If you are the root account holder, you should already have these permissions assigned to you. If not, you can click on **Add permissions** to search for your username and give yourself this permission, or it might be something that you have to ask your account administrator to do for you.

> **Important note**
> It's generally not advised to use the root account to spin up services via CDK or any other means directly. The root account should be used to give granular permissions to secondary users that can then deploy CDK stacks. We are, for now, skipping such steps to simplify the instructions.

Next, click on the **Security credentials** tab. Under **Access keys**, click **Create access key**. A popup should appear offering you the option to download the credentials as a CSV file or copy them directly from the popup. You can do whichever one of them you like. Once we use the credentials, don't forget to delete the CSV file on disk.

Now that we have the keys file downloaded from the AWS console, let's configure our local environment with these given keys. Run the following command:

```
$ aws configure --profile cdk
```

You will be prompted to enter the **AWS access key ID**. You can get this from the CSV file you downloaded or the value you copied over from the previous step. Once you have done this, press *Enter*.

Next, you will be prompted to enter the **AWS secret access key**. Just as in the previous step, copy this over and press *Enter*.

Now you will be prompted to enter a default Region name. We normally work within us-east-1 since it's one of the cheapest Regions (yes—AWS pricing differs per Region) and one of the first ones to receive new services and products. You can choose a Region near your geographical location if response promptness during the course of this book matters to you. us-east-1 has been selected—off we go.

Last, the tooling will ask you for a default output format; enter json.

Now, let's make sure you've set things up right. To do that, run the following command:

```
$ aws configure list --profile cdk
```

If you see an output like the one shown in the following screenshot, voilà! You're all set:

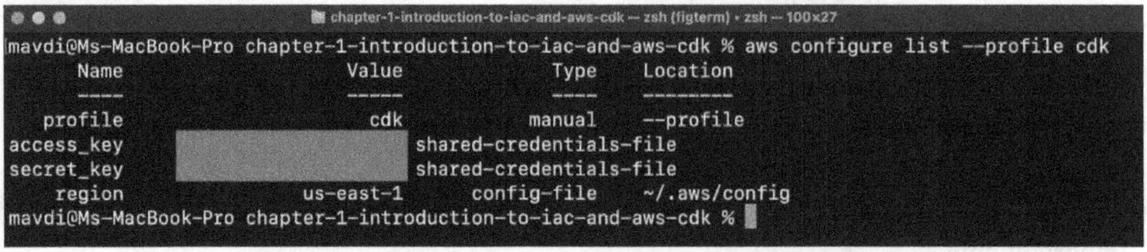

Figure 1.5 – Checking the configuration with the AWS CLI

> **Important note**
>
> As you see, even though the `list` command covers most of our keys and secrets, we're not taking any chances because even the last few characters could be used by a malicious actor for tracking. Keep your AWS keys as secret as your bank account details because if leaked, you could take a considerable financial hit.

Setting up the AWS CDK CLI

The AWS CDK CLI is a command-line tool that is installed as an npm package. If you're a JavaScript or TypeScript developer, you will have it already installed on your machine; otherwise, here is a link to the Node.js website to install this as well as Node's package manager, npm: `https://nodejs.org/en/download/`.

> **Note**
>
> Aim for the **Long-Term Support** (**LTS**) version of Node.js when downloading it. Generally, a good habit to hold as a developer is to use LTS versions of their tooling.

Once you have npm running, installing `cdk` on any operating system can be done by simply running the following command:

```
$ npm install -g aws-cdk@2.25.0
```

By running this, we are aiming for a specific version of the CLI, and we will also do the same with the library. CDK changes quite significantly between versions, therefore it's best for us to focus on a specific version of CDK. Once you've learned the ropes, you will be able to upgrade to a later version of it when available. At the time of writing this, 2.25.0 is the latest version of CDK.

Next, type in and run the following command:

```
$ cdk -h
```

If you see a bunch of commands like the ones seen in the following screenshot, you've done things right:

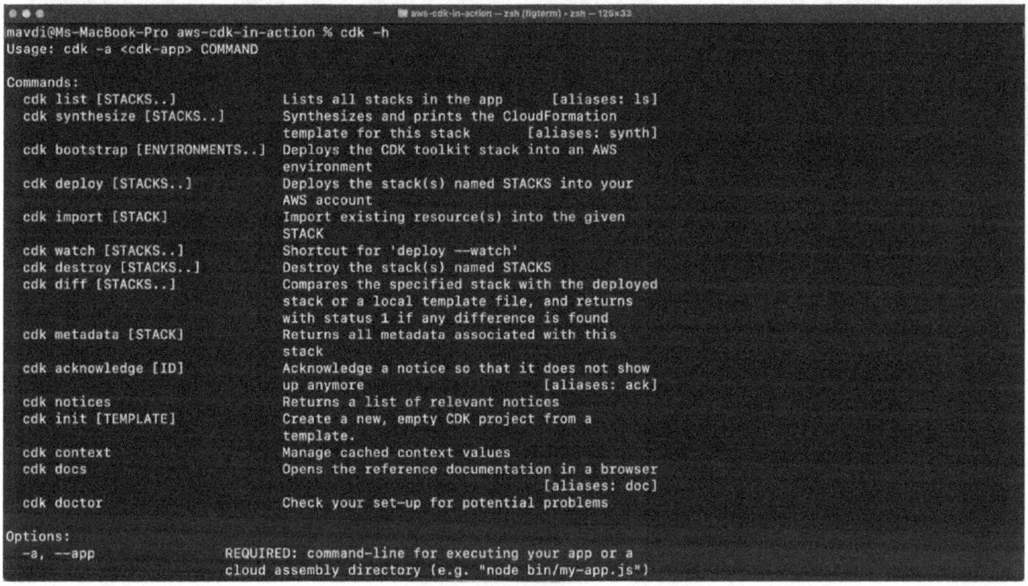

Figure 1.6 – Making sure the CDK toolset is installed correctly

Great! Now that we have the CDK CLI installed, let's go ahead and create our first CDK app.

Creating our first CDK app

Go ahead and create a directory on your workspace. In the example code provided, we've named it `chapter-1-introduction-to-iac-and-aws-cdk`. Next, bring up the CLI of your beloved operating system and run the following command inside the directory:

```
$ cdk init app --language typescript
```

A bunch of green text should appear while CDK does its magic. Once done, run the `ls` command to see the top-level folder structure, or even better, if you have the `tree` command installed, run it as such:

```
$ tree -I node_modules
```

The following is what you should be seeing:

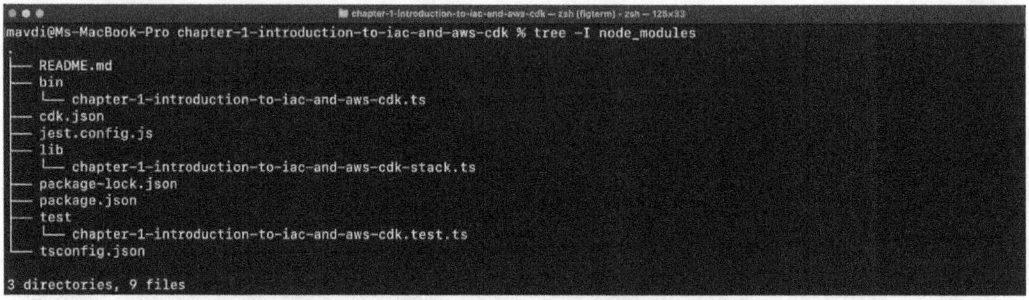

```
● ● ●                        📁 chapter-1-introduction-to-iac-and-aws-cdk — zsh (figterm) - zsh — 125×33
mavdi@Ms-MacBook-Pro chapter-1-introduction-to-iac-and-aws-cdk % tree -I node_modules

├── README.md
├── bin
│   └── chapter-1-introduction-to-iac-and-aws-cdk.ts
├── cdk.json
├── jest.config.js
├── lib
│   └── chapter-1-introduction-to-iac-and-aws-cdk-stack.ts
├── package-lock.json
├── package.json
├── test
│   └── chapter-1-introduction-to-iac-and-aws-cdk.test.ts
└── tsconfig.json

3 directories, 9 files
```

Figure 1.7 – Folder structure of our CDK app

If you don't have `tree` installed, you can install it with one of the following commands.

On Mac (using the Homebrew package manager), run this command:

```
brew install tree
```

On Linux (Ubuntu), run this command:

```
sudo apt install tree
```

> **Note**
> `tree` is already installed on Windows 10/11.

We've intentionally told the `tree` command to ignore `node_modules` since it's an abyss of directories we don't need to know about. At a high level, here are explanations of the files and directories:

- `README.md` is essentially a Markdown documentation file. You must've come across it.
- The `bin` directory is essentially where the top-level CDK app files reside:
  - `chapter-1-introduction-to-iac-and-aws-cdk.ts` is the file `cdk` created based on the name of the directory we ran the CLI `init` command in. This file contains the entry point of the application.
- `cdk.json` tells the toolkit how to run the application and passes along some further configurations.
- `jest.config.js` is the configuration file the `cdk` library uses to run local tests.
- `lib` contains all the goodies. All the constructs we will be creating for our project will be placed inside here, as outlined next:
  - `chapter-1-introduction-to-iac-and-aws-cdk-stack.ts` is one such construct or component. In this chapter, we will be spending most of our time here.

- You can safely ignore `package-lock.json` for now; it's what npm uses to keep track of specific versions of node libraries installed.

- `package.json` is the npm module's manifest information such as app, versions, dependencies, and so on.

- `test` is pretty self-explanatory. Our test files reside here, as detailed next:

  - `chapter-1-introduction-to-iac-and-aws-cdk.test.ts`: cdk has gone ahead and created a test file for us, urging us to test our application. We will do so in later chapters. For now, ignore it.

- `tsconfig.json` is where TypeScript-specific configuration is stored.

Now that we've learned about the files cdk has created, let's dive in and kick off our first AWS CDK app!

Creating a containerized web application in AWS CDK using Docker

In this step, we will keep things as simple as we can. The aim is to get you going with coding your first CDK app as quickly as you can. So, let's get started.

Open `lib/chapter-1-introduction-to-iac-and-aws-cdk-stack.ts` in your favorite editor. We're using Visual Studio Code. You will see the following code already present in the file. CDK has pretty much wired up everything for us and is ready to go:

```
import  * as cdk from 'aws-cdk-lib';
import { Construct } from 'constructs';
// import * as sqs from 'aws-cdk-lib/aws-sqs';

export class Chapter1IntroductionToIacAndAwsCdkStack extends cdk.Stack
{
  constructor(scope: Construct, id: string, props?: cdk.StackProps) {
    super(scope, id, props);

    // The code that defines your stack goes here

    // example resource
    // const queue = new sqs.Queue(this,
'Chapter1IntroductionToIacAndAwsCdkQueue', {
    //    visibilityTimeout: cdk.Duration.seconds(300)
    // });
  }
}
```

We could get rid of all the comments and start adding our code, but for the sake of simplicity, let's delete all that and paste the following code instead into the file:

```
import { Stack, StackProps } from 'aws-cdk-lib';
import { ContainerImage } from 'aws-cdk-lib/aws-ecs';
import { ApplicationLoadBalancedFargateService } from 'aws-cdk-lib/
aws-ecs-patterns';
import { Construct } from 'constructs';


export class Chapter1IntroductionToIacAndAwsCdkStack extends Stack {
  constructor(scope: Construct, id: string, props?: StackProps) {
    super(scope, id, props);

    new ApplicationLoadBalancedFargateService(this, 'MyWebServer', {
      taskImageOptions: {
        image: ContainerImage.fromRegistry('amazon/amazon-ecs-
sample'),
      },
      publicLoadBalancer: true
    });
  }
}
```

Alright—save the file, open up the command line, cd into the working directory (the root of our cdk application), and run the following:

```
$ cdk bootstrap --profile cdk
```

This command will bootstrap the AWS environment, meaning it will create all the necessary AWS resources on the cloud to make CDK work properly.

Then, run the following command:

```
$ cdk deploy --profile cdk
```

If you've followed the instructions correctly so far, you will see a nice green-text screen. Let's examine it:

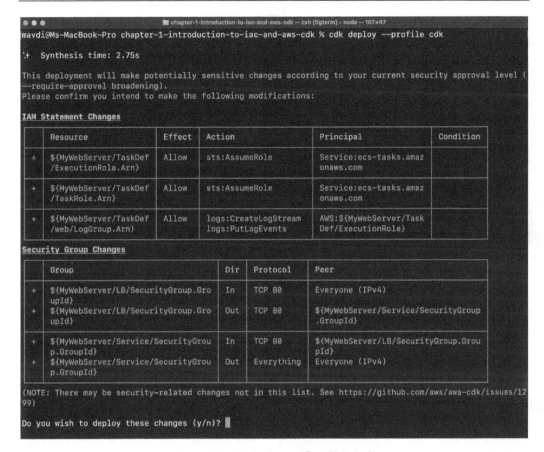

Figure 1.8 – Deploying our first CDK stack

The preceding screen is first warning you that potentially sensitive changes are about to occur with this deployment; this is due to the IAM roles the stack creates. It's fine for you to ignore this for now.

Next, it lists the names of AWS resources that are about to be created, categorized by service. The + sign and the green color of the text indicate that these resources are about to be added. Later, we will see resources that will be marked for destruction, which is logically indicated by a – sign at the beginning of the line and the color red.

Yes—you've earned it: go ahead and type y and press *Enter*.

> **Note**
>
> If you're in the middle of a stack being created and are about to take a break or attend to another matter, destroy the stack with `cdk destroy --profile cdk` to avoid unnecessary costs.

CDK will create an **AWS CloudFormation** template (we will cover this later in the chapter) and continue to deploy the stack. It will also keep you informed of the progress. If your user has got the necessary permissions, CDK will succeed in deploying the stack; else, it will fail and automatically roll back the changes.

The process is complete; we see a bunch of information being outputted. Under the **Outputs** section, you will see two links:

```
Outputs:
Chapter1IntroductionToIacAndAwsCdkStack.MyWebServerLoadBalancerDNSD1AFCC81 = Chapt-MyWeb-10RT1BQF3YPLM-9526
30869.us-east-1.elb.amazonaws.com
Chapter1IntroductionToIacAndAwsCdkStack.MyWebServerServiceURLB0ED50F6 = http://Chapt-MyWeb-10RT1BQF3YPLM-95
2630869.us-east-1.elb.amazonaws.com
Stack ARN:
arn:aws:cloudformation:us-east-1:531551740863:stack/Chapter1IntroductionToIacAndAwsCdkStack/bc928dd0-e105-1
1ec-90c0-0a909d956d03

 ✨  Total time: 271.42s
```

Figure 1.9 – CloudFormation stack output links

Copy either of the links into your web browser and navigate to it. You should see a page load up:

Figure 1.10 – The output of the Elastic Container Service (ECS) server

Nice! You've successfully deployed your first CDK application. If you now go back to your AWS dashboard and go to **ECS Service**, you will see a new cluster of services has been spun up on your behalf.

If you click on the cluster name, which should be `Chapter1IntroductionToIacAndAwsCdk Stack-{some other random text}`, then you will be able to see the service, tasks, metrics, logs, and all the rest. This is a containerized **ECS Fargate** application with a lot of built-in resiliency already spun up by you already. So many other good things are about to come.

Right—we've had our fun. Let's now go ahead and destroy the stack before we do anything else. To destroy the stack—you guessed it—run the following:

```
$ cdk destroy --profile cdk
```

Confirm by entering y. AWS CDK will now go ahead and destroy our favorite PHP stack. Perhaps it's for the best.

Going back to the dashboard and to the **ECS Service** root page, you will see that the cluster is no longer there. Neither is our simple PHP app. All gone. Clean.

Understanding the inner workings of AWS CDK

We hope that was fun. In the previous section, we mentioned **AWS CloudFormation** and how CDK outputs a **CloudFormation** template and then manages its life cycle.

According to AWS, CloudFormation is an IaC service (again, I'd argue with the code bit) that you can use to model, provision, and manage AWS services. In short, it's a **YAML** or **JSON** file with an AWS service definition of its properties and relationships.

Learning CloudFormation is outside the scope of this book, but it's useful for you to understand and read about it, to better debug your CDK applications. Let's take a brief look at a CloudFormation excerpt sample YAML configuration.

Here is how you set up a basic EC2 instance and open up the 22 port for SSH access. Reading YAML is straightforward, and if you look closely, you will be able to read the various components our CloudFormation configuration defines:

```yaml
Parameters:
  KeyName:
    Description: The EC2 Key Pair to allow SSH access to the instance
    Type: 'AWS::EC2::KeyPair::KeyName'
Resources:
  Ec2Instance:
    Type: 'AWS::EC2::Instance'
    Properties:
      SecurityGroups:
        - !Ref InstanceSecurityGroup
        - MyExistingSecurityGroup
      KeyName: !Ref KeyName
      ImageId: ami-7a11e213
  InstanceSecurityGroup:
    Type: 'AWS::EC2::SecurityGroup'
    Properties:
      GroupDescription: Enable SSH access via port 22
      SecurityGroupIngress:
        - IpProtocol: tcp
          FromPort: 22
          ToPort: 22
          CidrIp: 0.0.0.0/0
```

Well, CDK uses the same underlying mechanism. Working with AWS CloudFormation directly can be very daunting and complicated, even for relatively simple stacks. To prove this point, go to this chapter's CDK app root and run the following command:

```
$ cdk synth
```

You guessed it right—this gigantic abomination of a YAML output is the result of about 20 lines of CDK TypeScript code. CDK essentially compiles your code into a CloudFormation stack and manages the rest of the complexity of adding and removing various bits, linking resources together, and a ton of other things for you.

The amount of time that developers save is undeniably massive. The amount of confusion, mistakes, and painful trials and errors of CloudFormation or any other configuration-defined IaC tool that CDK eliminates makes CDK and the new set of similar tools such as **Pulumi** clear winners of the IaC race. Businesses that onboard CDK into their development practices will be able to deliver a lot more in a shorter amount of time.

Developers with CDK skills will be highly sought after. Welcome aboard—this is the future of software development on the cloud!

Summary

In this chapter, we took a brief look at problems and issues with the current IaC tools and methods. We introduced CDK, got a taste of what it is capable of, and understood some of the inner workings of CDK, as well as how potentially revolutionary it is.

In the next chapter, we will dive deeper into setting up our workspace, directory structures, and environment correctly. We will also explain the concept of constructs and stacks and get to know other important aspects of CDK.

2

A Starter Project and Core Concepts

At Westpoint, we've been invested in the **Cloud Development Kit** (**CDK**) platform right from the early days when we discovered the potential productivity gains. You will discover more about these potential gains later in the following chapters as we introduce new approaches to cloud solution architecture that this awesome new tool brings.

As a result of our constant curiosity toward CDK and later using it to deliver projects, we have come up with some good practices for organizing CDK-based applications. These practices revolve essentially around code organization and should not be considered complete, by any means, since we are still on this journey ourselves and we add things as we discover them. It works for us, so it might work for you. We would also love to hear of other ways CDK projects can be integrated into other modern workflows. So, if you have such suggestions, please reach out to us via any of our public GitHub repos.

In this chapter, we will learn about the following core topics:

- How to integrate CDK code into your project with the monorepo model
- Core CDK concepts such as **constructs**
- A basic overview of creating custom constructs
- Discover where to find community-maintained construct libraries

Technical requirements

Speaking of GitHub repos, you will be able to find our latest CDK starter project by following this link: `https://github.com/PacktPublishing/AWS-CDK-in-Practice/tree/main/chapter-2-starter-project-and-core-concepts/cdk-starter/infrastructure`.

The Code in Action video for this chapter can be viewed at: `https://packt.link/565U3`.

The CDK monorepo model

You are probably familiar with the concept of monorepos. The details are outside the scope of this book, but essentially, monorepos point to all code and assets of a certain project, client, or company being in a single GitHub repo as opposed to logically separating the code by, for example, separating the frontend and the backend of the code base into different repositories.

There are many upsides and downsides to using monorepos, and developers use them at varying levels for code organization, from keeping the entire company code within a repo to just storing project-specific code in a single repository. For example, at Westpoint, we like to keep each client's code in a separate monorepo within a separate GitHub organization. This way, we keep things secure and easier to configure.

But if you have different levels of developer access to different bits of your organization's code, it's best to use other methods. We aim to keep things simple to get started, as learning a new technology can be challenging enough. This way, we will be able to get going quickly.

More information on monorepos and their pros and cons can be found here: `https://semaphoreci.com/blog/what-is-monorepo`.

High-level separation of concepts

We start by creating directories for high-level application concepts. These are usually parts of the code that speak completely different languages in terms of code and APIs in such a way that not many components are shared between them.

Think about a **Next.js**-based web application that is deployed via CDK and **continuous integration/ continuous delivery** (**CI/CD**) on AWS services such as **CodeBuild** and **CodePipeline**. We want to keep the Next.js application and the infrastructure code separate. Hence, let's go ahead and create two root-level directories, `infrastructure` and `web`:

Figure 2.1 – Initial directory structure

Since this project has only one web application to be served, we've created a general web directory. Let's assume your organization manages multiple web applications that are logically and technologically separate. In that case, you would create a directory for each, for example, `sales-portal` and `main-website`.

We aim to manage the entire DevOps stack of all our web applications from the `infrastructure` directory. Let's run `cdk init` to kick off our CDK app:

```
$ cdk init app --language typescript
```

CDK concepts

We have initiated our CDK app, but before we go further, there are a few key CDK concepts that you should understand first. Let's delve into them one by one.

A CDK app

An AWS CDK app is an application written in a programming language, the most popular being TypeScript (see the language flag when we ran `cdk init`), which uses the standard CDK library alongside custom-written components to define an AWS-powered infrastructure.

Open up `bin/infrastructure.ts`. You will see the following line of code:

```
const app = new cdk.App();
```

The app constantly acts as the root of our CDK application. The reference to the app will be passed down to all our CDK stacks:

```
new InfrastructureStack(app, 'InfrastructureStack', {})
```

CDK stacks

Stacks act as high-level containers for constructs, then are used to define AWS services such as ECS, DynamoDB, and others. Looking at our starter repository so far, we will define only one stack – `WebStack`. It's recommended to have separate stacks for each AWS application you deploy. Following on the same multi-app example from the last section, we would have `SalesPortalStack` and `MainWebsiteStack` accordingly.

The current name of our main stack is `InfrastructureStack`. Let's change this now to `WebStack`. Let's also get rid of the comments. You will end up with the following in `infrastructure.ts`:

```
#!/usr/bin/env node
import 'source-map-support/register';
import * as cdk from 'aws-cdk-lib';
import { WebStack } from '../lib';

const app = new cdk.App();
new WebStack(app, 'WebStack', {});
```

Let's rename the `lib/infrastructure-stack.ts` file to just `index.ts`. We will keep a similar pattern going forward. This will make more sense once we learn about constructs. You can also go ahead and change the name of the stack to `WebStack` within the same file.

Good, let's now run `cdk synth` to verify that everything compiles. If you're getting a long YAML output with a lot of nested properties and values, you've done it right. It has successfully compiled:

```
              - us-east-2
        - Fn::Or:
            - Fn::Equals:
                - Ref: AWS::Region
                - us-west-1
            - Fn::Equals:
                - Ref: AWS::Region
                - us-west-2
Parameters:
  BootstrapVersion:
    Type: AWS::SSM::Parameter::Value<String>
    Default: /cdk-bootstrap/hnb659fds/version
    Description: Version of the CDK Bootstrap resources in this environment, automatically retrieved from SSM Para
eter Store. [cdk:skip]
Rules:
  CheckBootstrapVersion:
    Assertions:
      - Assert:
          Fn::Not:
            - Fn::Contains:
                - - "1"
                  - "2"
                  - "3"
                  - "4"
                  - "5"
                - Ref: BootstrapVersion
          AssertDescription: CDK bootstrap stack version 6 required. Please run 'cdk bootstrap' with a recent versio
of the CDK CLI.

mavdi@Ms-MacBook-Pro infrastructure %
```

Figure 2.2 – CDK translates your resources into CloudFormation templates

The `cdk synth` command is a validation of the soundness of the stack we are creating using CDK. To qualify for this check, the stack's TypeScript code must first compile and, additionally, check whether CDK tooling conducts should also be applied. Only then is the YAML output produced.

In the next chapter, we will look more closely at this YAML output you've just seen. For now, know that it's an AWS CloudFormation template file that CDK uses as its underlying deployment tooling. This output doesn't necessarily mean that your stack is assured to be deployed. There are further checks done at the stack's deployment time, and, for example, if a reference to another stack ends up invalid, then while the CDK app has compiled, the stack will still fail to be deployed.

CloudFormation is similar to HashiCorp's Terraform in that it checks AWS for the state of a certain stack at a time and given the new additions or removals, it creates a changeset that is then applied using the AWS API.

Now that we have learned about the concept of the CDK app, let's go one level deeper and look into CDK constructs.

The organization of constructs

Constructs are encapsulations of one or more AWS cloud components. They are the *classes* of the CDK world, and they are indeed defined as classes (so are apps and stacks). A construct can be as simple as a single resource declaration or might be very high-level abstractions such as an entire load-balanced web server that would accept an HTML file in its parameters and construct all the services needed to serve this file.

AWS CDK comes with a standard construct library. Detailed documentation of all constructs can be found here: `https://docs.aws.amazon.com/cdk/api/v2/docs/aws-construct-library.html`.

Almost all AWS resources, which is everything the CloudFormation template files define and deploy, are defined as constructs within the CDK library. Generally, CDK constructs can be divided into the following three categories:

1. **L1**: Short for *layer 1*, also called CFN resources, are low-level constructs that directly match CloudFormation declarations of AWS services one to one. They are all prefixed with `Cfn` (for example, `CfnBucket`, representing `AWS::S3::Bucket`).

2. **L2**: The next higher-level constructs of the CDK library are built on top of L1 constructs, and they come with boilerplates, defaults, Glue code, and helper functions for convenience. An example of this would be `s3.Bucket` (`https://docs.aws.amazon.com/cdk/api/v2/docs/aws-cdk-lib.aws_s3.Bucket.html`), which comes with helper functions such as `bucket.grantPublicAccess()` that internally take care of the complexities of the S3 bucket which is public.

3. **L3**: One level higher; these constructs are also called patterns. They are more complex best practice patterns that can be used to spin up complex structures of AWS services with sensible defaults. We used one such L3 construct in *Chapter 1*. `ApplicationLoadBalancedFargateService` is an L3 construct that sets up a load-balanced **Elastic Container Service** (**ECS**) Fargate container service.

Now, let's go back to our starter project. Dig into `infrastructure/lib/` and create a new directory named `constructs`. The constructs we will produce will look more like L3 constructs that come with the standard CDK library. Let's go ahead and produce such constructs.

Let's say, in this project, we want to create an S3 bucket for two separate environments. Let's do this by creating a new construct file. We can follow the same `index.ts` pattern that we've followed so far to create the following file:

`infrastructure/lib/constructs/S3Bucket/index.ts`

Open this file in your code editor and enter the following code. As a reminder, you can find all the chapter-specific code examples in the following GitHub repository:

```
import { RemovalPolicy } from "aws-cdk-lib";
import { Bucket } from "aws-cdk-lib/aws-s3";
import { Construct } from "constructs";
// This is an interface we use to pass the environment as a variable
to the construct
interface Props {
  environment: string;
}

export class S3Bucket extends Construct {
  public readonly bucket: Bucket;
  // Every construct needs to implement a constructor
  constructor(scope: Construct, id: string, props: Props) {
    super(scope, id);

    const bucketName = props.environment === 'production' ?
'bucket-s3' : 'bucket-s3-dev';

    this.bucket = new Bucket(scope, 'Bucket-S3', {
      bucketName,
      // When the stack is deleted, the bucket should be destroyed
      removalPolicy: RemovalPolicy.DESTROY,
      publicReadAccess: true,
    });
  }
}
```

In this relatively simple construct, we are creating an S3 bucket. Let's go over some of the details:

```
import { RemovalPolicy } from "aws-cdk-lib";
import { Bucket } from "aws-cdk-lib/aws-s3";
import { Construct } from "constructs";
```

The imports are self-explanatory. We are importing the Bucket L2 construct, the Construct class to be extended, and also the RemovalPolicy class, which we will explain shortly:

```
export class S3Bucket extends Construct {
```

Our custom construct extends the Construct base class. This base class represents the building blocks of CDK constructs:

```
constructor(scope: Construct, id: string, props: Props) {
```

All constructs, whether they're from the standard library or are custom-made, take three parameters:

- `scope`: Either a stack or another construct; this is the parent of the current construct. CDK uses this property to build a construct tree that forms the relationship between the root of the application and lower-level components. When writing CDK applications in TypeScript, usually the `this` value is passed down to this property.

- `id`: This represents a unique identifier within the scope. CDK uses this identifier to name CloudFormation (the YAML output we saw earlier) resources. Identifiers only need to be unique for a certain scope.

- `props`: These are the configuration properties passed down from the parent construct or stack. In this case, we are passing down the environment. We've also defined a TypeScript interface to facilitate this:

```
this.bucket = new Bucket(scope, 'Bucket-S3', {
    bucketName,
    // When the stack is deleted, the bucket should be
destroyed
    removalPolicy: RemovalPolicy.DESTROY,
    publicReadAccess: true,
});
```

In this section, we will initialize the bucket with certain defaults. One of these defaults is choosing the right name of the bucket based on the environment since AWS S3 bucket names are global, and they could clash if all buckets from all production or development environments had the same names.

We are also using `RemovalPolicy.DESTROY` to tell CDK that we would like to remove this bucket when the stack gets removed, no matter what's inside the bucket. AWS, by default, refuses to delete buckets that have files.

Another clever use of the environment property would be to remove the bucket regardless of the contents in the development environment and not delete it if, via environment variables, we detect that we are in production.

All right, now let's go ahead and initiate this construct in our main stack. Open the `infrastructure/lib/index.ts` file and replace the content with the following code:

```
import { Stack, StackProps } from 'aws-cdk-lib';
import { Construct } from 'constructs';
import { S3Bucket } from './constructs/S3Bucket';


export class WebStack extends Stack {

  constructor(scope: Construct, id: string, props?: StackProps) {
```

```
    super(scope, id, props);

    // The code that defines your stack goes here

    const bucket = new S3Bucket(this, 'MyRemovableBucket',
{environment: 'development'})
  }
}
```

Run cdk synth. The long YAML output points to things having gone well. Of course, we could also go ahead and deploy the application using cdk deploy, just like in the previous chapter:

```
$ cdk deploy --profile cdk
```

We can also subsequently destroy the stack using the following command:

```
$ cdk destroy --profile cdk
```

So, we have now essentially gone ahead and built our own L2 construct, which produces a removable S3 bucket and names it according to the environment passed down to it. This is a good example of encapsulating a standard CDK construct with certain defaults that our projects might need. In later chapters, we will be mostly focused on L3-type constructs.

Now that we've learned about various construct types, in the next section, we will look into where to find community-driven open source contracts to use and for development reference.

The Construct Hub

AWS has come up with a portal for sharing constructs in the same way npm functions for node packages. To look at AWS's attempt to consolidate community efforts to come up with reusable constructs, you can visit the following link: https://constructs.dev/. You will find a very useful set of constructs developed by AWS or community developers.

To view community constructs, visit the following URL: https://constructs.dev/ search?q=&langs=typescript&sort=downloadsDesc&offset=0&tags=community

Here, you will see a list of community-driven TypeScript-compatible constructs, as can be seen in the following screenshot:

Figure 2.3 – The Construct Hub Community section

Here you can find an incredible number of mostly L3 constructs that have simplified the process of creating complex stacks. Just a quick browse through, and you can see constructs that spin up Datadog monitoring services, ones that run standalone ECS Fargate tasks, and ones that create integrations between **Simple Queue Service (SQS)** and Slack notifications.

AWS has certain criteria that must be met before these constructs are published on the hub. Needless to say, despite the checks on the CDK developer's conduct, the constructs found here (especially community ones) should be examined thoroughly before being used in a production environment.

A link to the GitHub repo, alongside licensing information and other stats, is provided on each construct's page, as can be seen in the following screenshot:

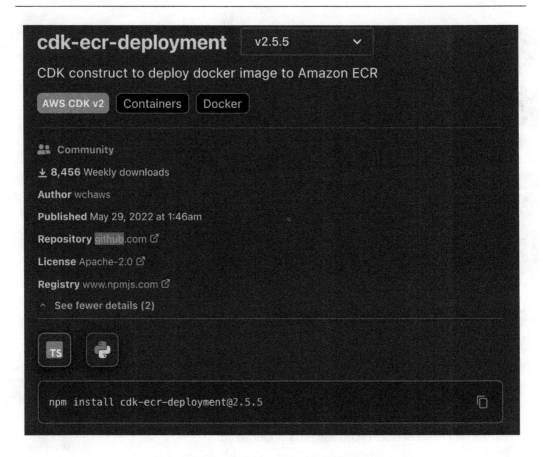

Figure 2.4 – Construct links, licensing, and installation details

These constructs can also be a great learning reference for general best practices when building CDK applications. We regularly read the source of new constructs to see the direction the community is taking and to find new ways of organizing the CDK code base.

Summary

In this chapter, we introduced a practical CDK starter project you can use when coding CDK. We also learned about some core CDK concepts such as apps, stacks, and constructs. Finally, we created a custom construct and used it in our stack.

In the next chapter, we will take this example further and create a web application that exposes a RESTful API. We will also finally look deeper into the CDK and AWS CloudFormation relationship and how understanding CloudFormation templates will help you when developing CDK applications.

Part 2: Practical Cloud Development with AWS CDK

Through a simple TODO application, in this part, we will cover the practical aspects of developing cloud applications using AWS **Cloud Development Kit (CDK)**. In *Chapter 3*, the book covers deploying Docker images on Amazon **Elastic Container Service (ECS)** and setting up DynamoDB tables and Amazon **Simple Storage Service (S3)** buckets via CDK. *Chapter 4* takes a deep dive into creating hosted zones in Route 53, validating **Domain Name System (DNS)** for **AWS Certificate Manager (ACM)** certificates, setting up a **Relational Database Service (RDS)** MySQL database, and safely storing secrets with Secrets Manager. In *Chapter 5*, the book teaches you how to set up GitHub and CodePipeline to work together to automate the deployment of stacks in multiple environments. *Chapter 6* focuses on testing CDK applications, including different types of tests, effective test-writing strategies, and best practices for integrating testing into the CDK development workflow. This part has the following chapters:

- *Chapter 3, Building a Full Stack Application with CDK*
- *Chapter 4, Complete Web Application Deployment with AWS CDK*
- *Chapter 5, Continuous Delivery with CDK-Powered Apps*
- *Chapter 6, Testing and Troubleshooting CDK Applications*

3

Building a Full Stack Application with CDK

In the previous chapter, we learned about the concept of monorepos and how they help you organize your CDK code alongside higher-level separations such as infrastructure, frontend, and backend code. As we mentioned earlier, this isn't a rule when working with CDK. We're sure there will be better ways discovered by the CDK development community as processes evolve and more developers get to use it in their projects. For now, this way of organizing your code is good enough, so let's see how it all comes to life in a practical fashion.

In this chapter, we are going to learn about the following main topics:

- Building a **Node.js** and **Express.js** backed API
- Creating a React application that integrates with the API
- Bringing it all to life with AWS CDK and using services such as **Elastic Container Service** and **DynamoDB**
- How CDK helps with building Docker images for ECS

You might already know how to build full stack applications with React and Node.js. If that's the case, just review the code and move ahead to the good bits about CDK.

Setting up and building the stack

If you haven't done so already, go ahead and clone the following repository: `https://github.com/PacktPublishing/AWS-CDK-in-Practice`.

The Code in Action video for this chapter can be viewed at: `https://packt.link/GZxqU`.

Heads up

In this chapter, we will also be using Docker tooling. You can find the installation setup for your operating system by visiting `https://docs.docker.com/get-docker/`.

You will find the code for this chapter in the relevantly named `chapter-3-building-a-full-stack-app-with-cdk` directory. Just like we did with the last chapter, we have separated the code into the following main directories:

- `infrastructure` will hold our CDK components
- `server` will contain the code for our Express.js-based API
- `web` is essentially a React-based frontend that hooks up to the API

If you dig into any of these directories, you will see that they each have their own README files, `package.json`, and various other relevant configurations for building these high-level components.

Frontend

In this chapter, we will be building a TODO application that is backed by an API and DynamoDB as a database. Let's go right ahead and deploy this TODO application and see how it works.

To do that, we need to create the frontend asset that CDK needs to push to AWS S3 to host the TODO application written in React. Run `cd web` and then run the following:

```
$ yarn
```

This will install the dependencies since it's not advised to push them into the repository. Next, run the following:

```
$ yarn build
```

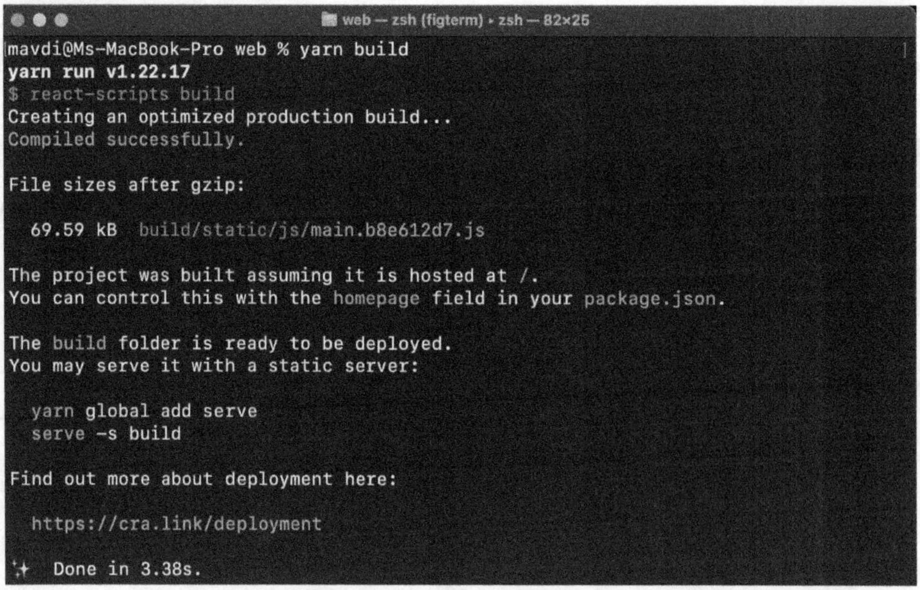

Figure 3.1 – The result of yarn build

If you're seeing output as shown in the screenshot, it's all gone well. This will create the `build` directory and place the compiled React assets within it.

Backend

With the frontend dealt with, open up the chapter's code in your favorite code editor and, in the terminal, run cd `server` to get into the `server` directory. Run the following command:

```
$ export AWS_PROFILE=cdk
```

Then run the following command:

```
$ export PORT=3001
```

This will essentially set both `AWS_PROFILE` and `PORT` as environmental variables for the opened terminal. Our backend will need to connect to DynamoDB, and we are going to be giving it access via the AWS **AccessKeyID** and **SecretAccessKey** that you provisioned for yourself. By exporting the AWS profile, you give CDK the directions to search for the credentials without having to create an `.env` file or, even worse, hardcode them somewhere else in the code. Note that this procedure is only needed to run the server code locally.

With that done, run the following:

```
$ yarn
```

Then run the following:

```
$ yarn dev
```

And you should see something like this in the terminal:

```
> yarn dev
yarn run v1.22.19
$ tsnd index.ts
[INFO] 17:32:40 ts-node-dev ver. 2.0.0 (using ts-node ver. 10.8.2, typescript ver. 3.9.10)
API listening on port 3001
```

Figure 3.2 – The result of the yarn dev infrastructure

Let's next look into spinning up the infrastructure section of the CDK app. Run cd `../ infrastructure` to get into the CDK portion of the code. Then, run the following:

```
$ yarn
```

Next, run the following:

```
$ cdk synth
```

This should present you with hundreds of lines of almost unreadable – unless you're into these kinds of things – YAML code. Despite that, this means things have gone well and our CDK application has been compiled. Mind you, this is the hundreds of lines of YAML, or similar **domain-specific language (DSL)**, code you avoid writing by using CDK.

Let's go ahead and deploy the app. Run the following:

```
$ cdk deploy --profile cdk
```

This will present you with a list of changes that CDK will make on your behalf categorized nicely by topic. You will then be asked whether you want to continue to deploy the app; go ahead and type y and press *Enter*.

You will notice that after the usual CDK-related asset uploading progress reports, a series of output messages regarding a Docker build appears:

```
infrastructure — zsh (figterm) · com.docker.cli — 82×25

#1 [internal] load build definition from Dockerfile
#1 sha256:90496fbe171997e1bd9461033b4243774839768a998b2d16de1c8c80e02a77cf
#1 transferring dockerfile: 163B done
#1 DONE 0.0s

#2 [internal] load .dockerignore
#2 sha256:b87714487970bdb9e72bc676f54a8ef9c078aa598abc10a43586c1cd38719732
#2 transferring context: 2B done
#2 DONE 0.0s

#3 [internal] load metadata for docker.io/library/node:lts
#3 sha256:2646f34a408a79d67d8ec4aee5fbe1e86dbb6088e389a0daa43d9ff3fb7376c1
#3 DONE 1.8s

#4 [1/4] FROM docker.io/library/node:lts@sha256:a13d2d2aec7f0dae18a52ca4d38b592e45
a45cc4456ffab82e5ff10d8a53d042
#4 sha256:23670c43de3ef2d47c67029dd3b113d64da0aad3ce46b21a6552551a179f29d0
#4 DONE 0.0s

#6 [internal] load build context
#6 sha256:9044b0a00084c22d9328b46217a34e1b97008d464e77e890b7c980b660fe2c2b
#6 transferring context: 1.98kB 4.4s
#6 transferring context: 77.63kB 4.9s done
#6 DONE 5.0s
```

Figure 3.3 – CDK building the Docker image

This is yet more CDK magic. The toolset realizes that you are using Elastic **Container Service** (**ECS**) and pointing to a Docker file in the code base. It then grabs this Dockerfile, builds the image on your behalf, and then even logs it to **Elastic Container Registry** (**ECR**) and pushes the image! What a time to be alive.

Wait for the stack to deploy. You will receive an output URL that should look something like the following:

Figure 3.4 – CDK output of the API URL

Open up your web browser and navigate to the API URL output. Here is what you should see:

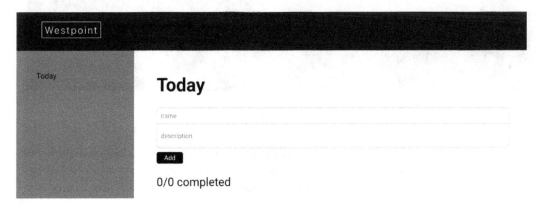

Figure 3.5 – Our API has come to life

Lovely. Next, navigate to the frontend URL output value; make sure you add `http://` to the beginning and `/index.html` to the end of it. Let's see what that gives us:

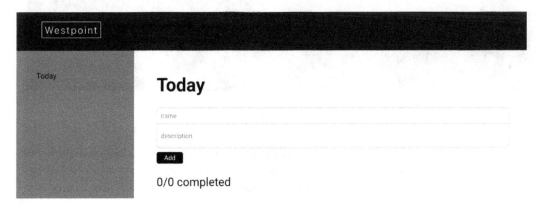

Figure 3.6 – Frontend being served from S3

Go ahead and enter the values to create a TODO item in the list. Before you click **Add**, open up the inspector in your browser (usually activated by right-clicking on the page and selecting **Inspect**, or by pressing *F12*) and go to the **Network** tab. Now, click the **Add** button of the TODO app:

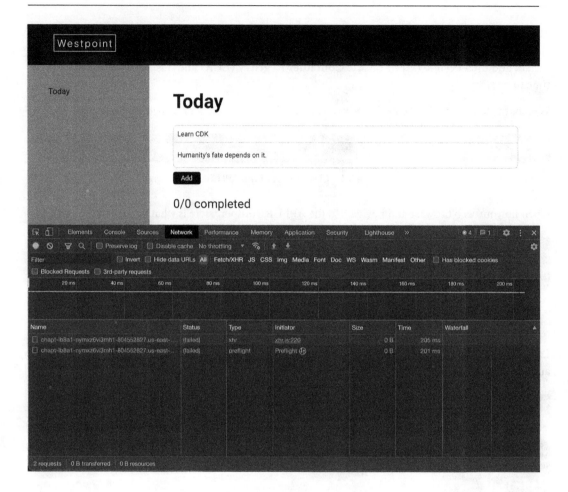

Figure 3.7 – Frontend API calls fail

As seen, our frontend is not able to make API calls to the backend. Now, before you toss the book (or whatever device you're using to read the e-book) out the window, we can explain what's happening. For now, we will give you the quick fix, while in the next chapter, we will learn how to properly address this.

Fixing the frontend code

The frontend is a standard React application. The application entry point is index.ts, which is located under the ./web/src directory. This, in turn, loads the App component. The App component holds the Header, SideBar, and Main components of our application. You can examine the rest of the code by yourself, but the bits we are interested in are within the Main component, which is located at `/web/src/components/Main/index.ts`. Open the file in your text editor and let's examine what is happening with our frontend not being able to make the calls to our API. See the following extract of the file:

```
...
{ MainContainer } from './styles';

/* ----------
 * Add backend URL provided by the cdk deploy here!
 * ---------- */
const backend_url = 'http://localhost:3333/';

export const Main: React.FC = () => {
  /* ----------
   * States
   * ---------- */
  const [todos, setTodos] = useState<Interfaces.Todo[]>([]);

  useEffect(() => {
    const fetchTodos = async () => {
      const response = await axios.get(backend_url);

      setTodos(response.data.todos);
    };

    fetchTodos();
  }, []);
...
```

The problem is immediately visible. On closer examination, we see that as soon as the component loads, we are making an API call with the `axios` library:

```
const response = await axios.get(backend_url);
```

However, the `backend_url` constant used by the library to fetch the TODO list is set to `localhost`:

```
const backend_url = 'http://localhost:3333/';
```

We need to change this to the backend URL set by CDK, which was printed in the output when the stack was fully deployed. Copy that URL and replace the `localhost` URL with it. It should look something like the following; don't forget to add `http://`:

```
const backend_url = 'http://Chapt-LB8A1-ZXACMP3PMFL0-1880097408.
us-east-1.elb.amazonaws.com';
```

Not the right way

Manually copying the output from one command and saving it into a file to deploy a stack is definitely not the right way to deploy. We will cover how best to do this in *Chapter 4*.

Save the file and build the frontend again by navigating to the /web directory in your terminal and running the following command:

```
$ yarn build
```

Next, go to the /infrastructure directory in your terminal window and redeploy the stack by running the following command:

```
$ cdk deploy --profile cdk
```

Observation

You will notice that the stack takes less time to deploy. This is because CDK and the underlying AWS CloudFormation are smart enough to only change the bits they have to modify, which, in this case, is the frontend code.

Wait for the stack to deploy and copy the new FrontendURL, add http:// to the beginning and /index.html to the end of it, and navigate to the page via your web browser. Add the TODO item again and you will see that it makes the correct API call:

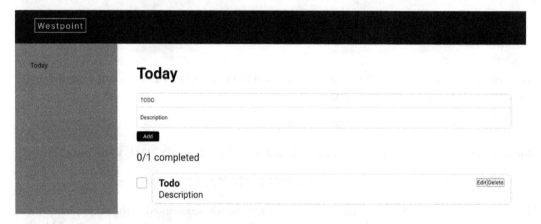

Figure 3.8 – TODO item added successfully

Let's see how our DynamoDB database is doing. Log in to your AWS console and find DynamoDB in the list of services. Click on the **Tables** button on the menu on the left. Click on main_table next and, finally, the big orange button in the top right, which says **Explore table items**.

Figure 3.9 – Data ends up in a DynamoDB table

We've done it. Our application works. Let's now examine the CDK code and see what made all of this come to life.

Examining the CDK code

As you've already guessed, the CDK code resides within the /infrastructure directory, with the CDK entry point being the /infrastructure/bin/chapter-3.ts file:

```
#!/usr/bin/env node
import 'source-map-support/register';
import * as cdk from 'aws-cdk-lib';

import { Chapter3Stack } from '../lib/chapter-3-stack';

const app = new cdk.App();

new Chapter3Stack(app, 'Chapter3Stack', {});
```

Nothing fancy is going on here. We are just telling it to load up the root stack of the CDK app, which is Chapter3Stack. Open /infrastructure/lib/chapter-3-stack.ts in your code editor:

```
export class Chapter3Stack extends Stack {
  public readonly dynamodb: Dynamodb;

  public readonly s3: S3;

  public readonly ecs: ECS;

  constructor(scope: Construct, id: string, props?: StackProps) {
    super(scope, id, props);

    this.dynamodb = new Dynamodb(this, 'Dynamodb');
```

```
    this.s3 = new S3(this, 'S3');

    this.ecs = new ECS(this, 'ECS', {
      dynamodb: this.dynamodb,
    });
  }
}
```

Our stack consists of three main constructs – a DynamoDB database, an S3 bucket to hold our frontend files, and an ECS service that hosts our backend. Note here that once we have instantiated the DynamoDB table, the ECS service gets a reference to it. This makes sense since our backend needs to know where to reach the DynamoDB table.

Let's now examine each of the constructs.

DynamoDB table

Spinning up a DynamoDB table with CDK is very straightforward. All we have to do is instantiate the `Table` class from the `aws-cdk-lib/aws-dynamodb` package. Let's see how this is done.

Open the `/infrastructure/lib/constructs/Dynamodb.ts` file:

```
constructor(scope: Construct, id: string) {
    super(scope, id);

    this.main_table = new Table(scope, 'MainTable', {
      partitionKey: {
        name: 'partition_key',
        type: AttributeType.STRING,
      },
      sortKey: {
        name: 'sort_key',
        type: AttributeType.STRING,
      },
      tableName: 'main_table',
      billingMode: BillingMode.PAY_PER_REQUEST,
      removalPolicy: RemovalPolicy.DESTROY
    });
  }
```

We could author an entire book on DynamoDB alone, so we won't go in depth into the various configurations and properties of DynamoDB tables. You can find detailed documentation of DynamoDB CDK level 2 constructs here: `https://docs.aws.amazon.com/cdk/api/v1/docs/aws-dynamodb-readme.html`.

And if you want to dive in deeper to learn more about the awesome power of DynamoDB, more information can be found by visiting the following link: `https://docs.aws.amazon.com/amazondynamodb/latest/developerguide/Introduction.html`.

We have kept this as simple as we can here. `partitionKey` and `sortKey` are essentially indexing attributes of a DynamoDB database. The rest are as follows:

- `tableName`: Self-explanatory – it's the name of the table you saw in the AWS console.
- `billingMode`: DynamoDB has two types of billing modes – **PROVISIONED** and **PAY_PER_REQUEST**. Each is useful in certain circumstances. More information can be found here: `https://docs.aws.amazon.com/amazondynamodb/latest/APIReference/API_BillingModeSummary.html`.
- `removalPolicy`: Since this is a test application and we don't care about our data, we are explicitly instructing AWS to delete the table and all its content when we destroy our stack. You should not be doing this for production applications. When we learn about environments, we will see that we can set this flag as `true` for the development environment and `false` for the production environment of the same applications.

Next, let's take a look at our S3 construct.

S3 bucket

As mentioned earlier, our frontend code is practically hosted by an S3 bucket. This is because the frontend React application is a browser-rendered **single-page application** (**SPA**) and S3 has the capability to act as a content server for our files.

Open the `/infrastructure/lib/constructs/S3.ts` file in your code editor:

```
this.web_bucket = new Bucket(scope, 'WebBucket', {
    bucketName: `chapter-3-web-bucket-${uuidv4()}`,
    websiteIndexDocument: 'index.html',
    websiteErrorDocument: 'index.html',
    publicReadAccess: true,
    removalPolicy: RemovalPolicy.DESTROY,
    autoDeleteObjects: true,
});
```

Here, we are declaring the bucket, naming it, and letting AWS know that `index.html` acts as both `websiteIndexDocument` and `websiteErrorDocument`. In your production applications, you would ideally have `404.html` defined for `websiteErrorDocument`.

We are setting the bucket files to be publicly readable by setting `publicReadAccess` to `true`. Also, just like our DynamoDB table, setting the removal policy to `DESTROY` will remove the bucket once the stack is destroyed.

One additional bucket removal configuration here is `autoDeleteObjects`. This is needed since AWS will otherwise still refuse to remove the bucket when the stack is destroyed if there are files saved in it. This essentially gives AWS the heads up that we don't much care about the contents of the bucket, and it can safely remove those before also removing the bucket.

Next, we are telling CDK to grab all the files in the `./web/build` directory, which contains our frontend build files, and place them in the bucket:

```
this.web_bucket_deployment = new BucketDeployment(scope,
'WebBucketDeployment', {
    sources: [Source.asset(resolve(__dirname, '..', '..', '..',
'web', 'build'))],
    destinationBucket: this.web_bucket,
  });
```

CDK will take care of the complexities of uploading these files into the bucket during the deployment process. Thanks, CDK!

And finally, remember the output URL for the frontend? Well, that's what the following line of code generates:

```
new CfnOutput(scope, 'FrontendURL', { value: this.web_bucket.
bucketDomainName });
```

With the S3 bucket taken care of, let's look at how the backend is hosted via ECS.

ECS setup

Remember how, in the first chapter, we used a level 3 construct to deploy a Fargate-powered container in ECS? Here, we will keep the Fargate container, so we don't need to worry about managing underlying EC2 instances with an **auto scaling group**, which means a swarm of EC2 machines configured to run ECS-powered containers.

Open up the file located at `./infrastructure/lib/constructs/ECS.ts` in your code editor. Let's examine the code:

```
this.vpc = new Vpc(scope, 'Vpc', { maxAzs: 2 });
```

Since Fargate will be spinning up EC2 machines that in turn run our ECS containers, we will need a virtually isolated private cloud, a **VPC**. It's an environment with a network isolated from the rest of the customers of AWS for you to safely set up virtual machines or do all kinds of networking magic. That's all you need to know about that for now:

```
    this.cluster = new Cluster(scope, 'EcsCluster', { vpc: this.vpc
});

    this.cluster.addCapacity('DefaultAutoScalingGroup', {
      instanceType: new InstanceType('t2.micro'),
    });
```

Next, we are setting up an ECS cluster that holds a group of ECS services (we will only have one for now):

```
    this.task_definition = new FargateTaskDefinition(scope,
'TaskDefinition');

    this.container = this.task_definition.addContainer('Express', {
        image: ContainerImage.fromAsset(resolve(__dirname, '..', '..',
'..', 'server')),
        memoryLimitMiB: 256,
        logging: LogDriver.awsLogs({ streamPrefix: 'chapter3' }),
    });
```

This section creates the ECS task definition. A task definition contains information about what the ECS service needs to run. In this example, we are pointing to the Dockerfile located at `./server/Dockerfile`. Define how much memory should be assigned to the task and also ask AWS to kindly keep hold of the logs of the application. As seen during the deployment, CDK builds the image on your behalf and deals with all the necessary steps to upload the image to ECR and make it available for the ECS task definition:

```
this.container.addPortMappings({
    containerPort: 80,
    protocol: Protocol.TCP,
});

    this.service = new FargateService(scope, 'Service', {
      cluster: this.cluster,
      taskDefinition: this.task_definition,
    });

    this.load_balancer = new ApplicationLoadBalancer(scope, 'LB', {
      vpc: this.vpc,
      internetFacing: true,
    });

    this.listener = this.load_balancer.addListener('PublicListener', {
port: 80, open: true });

    this.listener.addTargets('ECS', {
      port: 80,
      targets: [
        this.service.loadBalancerTarget({
          containerName: 'Express',
          containerPort: 80,
        }),
      ],
      healthCheck: {
```

```
        interval: Duration.seconds(60),
        path: '/healthcheck',
        timeout: Duration.seconds(5),
      },
    });
```

Bear with me, we are nearly there. The preceding code essentially sets up all the port mapping and the load balancer for our backend. It tells the load balancer that it should forward any traffic it receives on port 80 and hand it over to our ECS service. It also indicates to the load balancer that it can check whether the service is up by periodically calling the /healthcheck endpoint of our backend application.

Finally, just like FrontendURL, we are writing the load balancer URL to the output as BackendURL:

```
new CfnOutput(scope, 'BackendURL', { value: this.load_balancer.
loadBalancerDnsName });
```

We are also granting all the necessary permissions so that the API can perform the desired actions on DynamoDB, in this case, read and write permissions:

```
props.dynamodb.main_table.grantReadWriteData(this.task_definition.
taskRole);
```

There is various frontend and backend code written for this chapter that we haven't covered here since learning more about Node.js, Express, and React is outside the scope of this book. If you're familiar with these technologies, go ahead and modify the frontend and backend code, redeploy the stack, and experiment with the changes. If this is all new to you, it acts as a good reference for learning.

Once you're done with this chapter, don't forget to destroy the stack:

```
$ cdk destroy --profile cdk
```

Summary

You now know how to deploy any Docker-based image on ECS using the power of CDK. Not only that but you also know how to set up DynamoDB tables and serve static HTML files and SPAs via S3 buckets.

In the next chapter, we will address a few of the problems of our TODO application. We will learn how to set up **continuous integration/continuous delivery (CI/CD)** and we will also secure our application by generating TLS certificates for our load balancer. All done via CDK.

Complete Web Application Deployment with AWS CDK

In the previous chapter, we learned how to deploy a simple web service with AWS CDK by utilizing AWS ECS for hosting and DynamoDB as a database. We built a TODO application that created an API and a frontend React application. While we built a working full stack cloud application, there were a few problems with our deployment:

- Neither the frontend nor the backend was secured via TLS

- We had to copy over the API URL into the frontend and redeploy the stack to make things work

- The frontend code was being directly served from S3 with no distributed content delivery mechanism

In addition to that, having DynamoDB as a database is somewhat cheat code. Not every web application can switch databases overnight. Perhaps you want to move the IaC part of an existing application to AWS CDK. What if this application's database is in MySQL?

We will attempt to address these points in this chapter. In summary, in this chapter, we will learn about the following:

- Setting up DNS for both frontend and backend URLs using **Route 53**

- Creating an AWS **RDS**-backed MySQL database with CDK and seeding data

- Securing these endpoints with AWS **ACM**-provided TLS certificates

- Setting up a **CloudFront** distribution for our frontend assets

Let's get started.

Technical requirements

The source code for this chapter can be found at `https://github.com/PacktPublishing/AWS-CDK-in-Practice/tree/main/chapter-4-complete-web-application-deployment-with-aws-cdk`.

The Code in Action video for this chapter can be viewed at: `https://packt.link/PbobK`.

Since this chapter is a continuation of the previous one, the directory structure is the same but with a few modifications to the services, which we will explore in this chapter.

You will also need a domain name that is not in use. This will allow AWS to take care of DNS records of the domain, a practice you have to go through to be able to create complete and solid web applications that are backed properly by TLS certificates issued to those domains.

DNS with Route 53

Let's say you own the domain name `example.com` and you would like to assign the following two DNS records to the frontend and backend portions of our TODO application:

- `frontend.example.com`
- `backend.example.com`

To do this, you would have to let AWS's DNS routing service, Route 53, take care of your domain's DNS records. You potentially have a domain name that you've parked and that's not currently used. If you do, great. If not, you can always register the cheapest possible domain name to complete this section of this book.

This is important because you will never release a web application with randomly assigned load balancer URLs, the likes of which we saw in the previous chapter. AWS Route 53 also has a domain registration service. Let's look at how it works:

1. Sign in to the AWS console and, in the top search bar, type in `Route 53`. Click on the result.
2. From the left-hand panel, click **Registered domains**.
3. Press the **Register Domain** button at the top.

You will see a screen where you can type in a domain name of your liking with the domain extension `.com` preselected. At the time of writing, this costs $12:

To register a domain name, start by finding one that's available. Enter the first part of the name (such as example in example.com), choose an extensi click Check. We'll tell you whether it's available and whether you can get it with other extensions. Learn more.

Figure 4.1 – Route 53 domain search page

The simplest way of completing this chapter is by choosing the `.click` domain extension, which only costs $3, and registering a domain. It might also come in handy in your future testing or perhaps when you're launching a start-up. Now that you know CDK, you can launch it quickly.

Wow, look at this! We found an awesome `cdkbook.click` domain. Let's go ahead and add it to the cart and go to the checkout. Business expenses be damned!

Availability for 'cdkbook.click'

Domain Name		Status	Price /1 Year	Action
cdkbook.click	✓	Available	$3.00	Add to cart

Figure 4.2 – Buying an affordable domain

Go ahead and give our internet overlords your address details to register the domain. Once registered, click **Go Back to Domains**. You will see the domain name waiting under **Pending requests**.

> **Apologies**
>
> We're sorry to make you do all this. As you might have noticed, `.click` domain extensions don't have a WHOIS privacy option. We guess we get what we pay for. Registering a domain via AWS is the simplest way, but you can also have AWS manage the DNS for another domain you might have. Details can be found at `https://docs.aws.amazon.com/Route53/latest/DeveloperGuide/migrate-dns-domain-in-use.html`.

While we wait for our pending domain to brew, let's go ahead and prepare the rest of this project.

Configuring the project

Open this chapter's code in your editor of choice. Just like in the previous chapters, we have divided the code into `infrastructure`, `server`, and `web` directories.

First, go into the `root` directory and open the `config.json` file. In it, you will see an object with a property for `domain_name`. Change the value to the domain name you just purchased and verified through Route 53, as follows:

```
1  {
2      "domain_name": "cdkbook.click",
3      "backend_subdomain": "backend-cdk-book",
4      "frontend_subdomain": "frontend-cdk-book"
5  }
6
```

Figure 4.3 – Adding the domain name to the CDK configuration

Next, go ahead and build the frontend for our application so that, just like in the previous chapter, CDK can upload the `build` directory contents to S3. Go to the `web` directory and, in the terminal, run the following command:

```
$ yarn
```

Follow up the preceding command with the following:

```
$ yarn build
```

Change directories to the `infrastructure` folder and run the following command in the terminal to install all of the dependencies as well:

```
$ yarn
```

Great – we have now bought and configured our domain name through the Route 53 domain registration dashboard. Now, let's go ahead and complete the last few steps so that we have a functioning DNS configuration.

Completing the AWS console journey

Once you've asked to register a domain and completed the necessary steps, such as verifying your email, you will receive another email from AWS confirming the registration. AWS will go ahead and create a **hosted zone** for the domain. Hosted zones are essentially the way AWS manages internal and public-facing DNS records.

Next, we must create a TLS certificate. We don't want pesky prying eyes on the contents of the traffic between our web app and the user's browser. To do this, go ahead and type `Certificate Manager` in the top AWS console search box. Click on the result. Click on **Request**, select **Request a public certificate**, and click **Next**.

In the **Fully qualified domain name** section, enter your domain name. In our case, that's `cdkbook. click`. Then, click the **Add another name to this certificate** button and enter `*.<your domain name.extension>`; in our case, that's `*.cdkbook.click`. Set the validation method to DNS validation and click **Request**. Once requested, find the certificate from the **Certificate Manage** console and look at its details:

Domain	Status	Renewal status	Type	CNAME name	CNAME value
cdkbook.click	⏱ Pending validation	-	CNAME	⧉ _22bd52befe3d4e6de8848a53037 9327d.cdkbook.click.	⧉ _9f0198372191c4 2bef3.zxwlrjxpwn. validations.aws.
*.cdkbook.click	⏱ Pending validation	-	CNAME	⧉ _22bd52befe3d4e6de8848a53037 9327d.cdkbook.click.	⧉ _9f0198372191c4 2bef3.zxwlrjxpwn. validations.aws.

Figure 4.4 – Checking the domain's validation status when requesting ACM certificates

Now, if you go back to the list of certificates, you will see the certificate marked as issued. Your Jedi training is complete.

Now, we have to go back to our hosted zone and enter the details ACM needs to verify the domain name ownership so that it can issue a certificate for us. Necessary bureaucracy, we're afraid. Copy the CNAME name and CNAME values for each of the names, as shown in the preceding screenshot.

> **Heads up**
>
> The CNAME record's name and value should be the same for both the domain and the subdomain. So, in reality, you need to create one CNAME record. Also, when copying the CNAME name into the **Route 53 Hosted zone** setting, don't copy the domain name itself at the end of the string. That bit is added by default by the Route 53 UI.

Go back to the Route 53 dashboard and click on **Hosted zones**. Find and click on your domain name. Click on **Create a Record** and enter the values you copied previously for the CNAME name and value. Leave everything else as is and click **Create records**. Give it five minutes and then check the ACM UI. It should have marked the status as **Success**:

Domain	Status	Renewal status	Type	CNAME name	CNAME value
cdkbook.click	⊘ Success	-	CNAME	⬐ _22bd52befe3d4e6de8848a53037 9327d.cdkbook.click.	⬐ _9f0198372191c4 2bef3.zxwlrjxpwn. validations.aws.
*.cdkbook.click	⊘ Success	-	CNAME	⬐ _22bd52befe3d4e6de8848a53037 9327d.cdkbook.click.	⬐ _9f0198372191c4 2bef3.zxwlrjxpwn. validations.aws.

Figure 4.5 – Successful validation of DNS in ACM

Now we have configured everything, let's go back to the infrastructure directory in our project and run the following infamous command in the terminal:

```
$ cdk deploy --profile cdk
```

You will see a long list of resources that CDK is about to create. Enter y in the terminal prompt and press *Enter*.

Just like in the previous chapter, AWS CDK will now go ahead and deploy the stack. Now, you will be able to follow the steps either through the terminal or on the AWS console by navigating to the CloudFormation service:

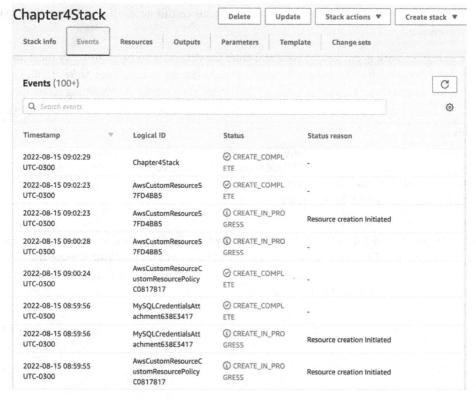

Figure 4.6 – Following the stack deployment process in CloudFormation

Once the stack has been deployed, go ahead and visit the frontend URL for your domain, which will be at `frontend.<your_domain_name>.<extension>`. In our case, it'll be `frontend.cdkbook.click`.

Hopefully, you will see the frontend TODO application from the previous chapter with a TODO item already added:

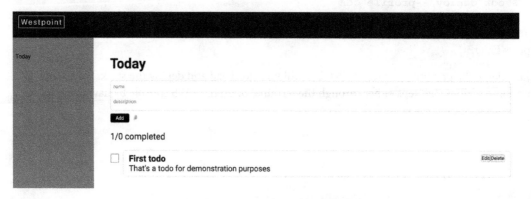

Figure 4.7 – TODO application deployed and ready

There you have it – your domain has a valid TLS certificate serving a CDK powered frontend React application with a backend that is also protected by TLS! This is the proper way of doing this. Now, let's look at a couple of other goodies we've added to cover more day-to-day scenarios.

MySQL powered by AWS RDS

In this chapter, as well as teaching you about Route 53, DNS, and certificates, we've also changed our main database to MySQL hosted on AWS's **Relational Database Service** (**RDS**).

Let's assume that you currently run a website with a MySQL database, and this book has inspired you to move to AWS and write your infrastructure code with AWS CDK. At some point, you are going to have to do a database migration. There are plenty of ways to do this with AWS, with the most complete solution being with **AWS Database Migration Service**.

Keeping with the theme of simplicity and giving you the tools to tackle infrastructure problems with AWS CDK, let's go with the simplest scenario, which would essentially be an exported .sql from your current database. That said, our plan is pretty evolved and advanced and should cover most types of database migrations.

If you look in the `infrastructure/lib/constructs/RDS/init` directory, you will see one such file named `script.sql`. When you open the file, you will see the following content:

```sql
USE todolist;

CREATE TABLE IF NOT EXISTS Todolist (
    id INT PRIMARY KEY NOT NULL AUTO_INCREMENT,
    todo_name VARCHAR(255),
    todo_description VARCHAR(255),
    todo_completed BOOLEAN
);

INSERT INTO
    Todolist (
        `todo_name`,
        `todo_description`,
        `todo_completed`
    )
VALUES
    (
        'First todo',
        'That''s a todo for demonstration purposes',
        true
    );
```

This file creates a SQL table named Todolist in the todolist database and defines its properties. Then, it inserts one Todolist item into the database.

> **Same principles**
>
> Your .sql file could be many GBs larger but the principles are the same. So long as you have this .sql file, you can use this method to populate your RDS database with it.

Our plan can be described with the following steps:

1. Create the RDS MySQL instance using AWS CDK.

2. Once created, tie in the CloudFormation step completion of the database to a **custom resource**.

3. This custom resource, in turn, triggers a custom Docker image **AWS Lambda** function.

4. The Lambda function connects to the database and populates it using the script.sql file.

> **New concepts**
>
> AWS Lambda and custom resource are new concepts in this book. Don't worry – we will briefly explain what they do shortly.

First, let's look at how the RDS instance is spun up. Navigate to /infrastructure/lib/constructs/RDS and open the index.ts file. Aside from the class definitions and other pleasantries, there is a database secrets management portion, which we will visit later in this chapter. For now, here is how the database resource is created:

```
this.instance = new rds.DatabaseInstance(scope, 'MySQL-RDS-Instance',
{
    credentials: rds.Credentials.fromSecret(this.credentials),
    databaseName: 'todolist',
    engine: rds.DatabaseInstanceEngine.mysql({
      version: rds.MysqlEngineVersion.VER_8_0_28,
    }),
    instanceIdentifier: instance_id,
    instanceType: ec2.InstanceType.of(
      ec2.InstanceClass.T2,
      ec2.InstanceSize.SMALL,
    ),
    port: 3306,
    publiclyAccessible: false,
    vpc: props.vpc,
    vpcSubnets: {
      onePerAz: true,
      subnetType: ec2.SubnetType.PRIVATE_ISOLATED,
```

```
    },
  });
```

Here is what's happening:

- We are creating a MySQL RDS database instance with the MySQL engine version `8.0.28`.

- The name of the database will be `todolist`.

- The database will be hosted by an EC2 machine of the T2 type with a `Small` instance size.

- We are defining ports and their accessibility within the VPC.

- We are using a VPC that is passed down higher on the stack and asking RDS to spin up the EC2 instance within a private subnet.

Next, we have to define the triggers for RDS resource completion. This can be achieved by using a CDK construct we found on the Construct Hub (`https://constructs.dev/`) website. We've examined the code and it looks good. The following figure shows a diagram of how it works, but we highly recommend that you read about it by going to `https://github.com/aws-samples/amazon-rds-init-cdk`:

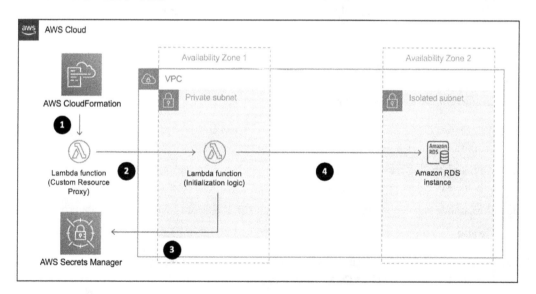

1. Trigger custom resource creation or update
2. The "Custom Resource Proxy" lambda invokes Lambda function with initialization logic.
3. Amazon RDS instance credentials are pulled from AWS Secrets Manager at runtime.
4. Initialization logic is executed on Amazon RDS instance.

Figure 4.8 – Architectural diagram of how Amazon RDS init works

Here is how we are initializing the custom (level 3) construct:

```
const initializer = new CDKResourceInitializer(scope, 'MyRdsInit', {
    config: {
      credentials_secret_name,
    },
    function_log_retention: RetentionDays.FIVE_MONTHS,
    function_code: DockerImageCode.fromImageAsset(`${__dirname}/
init`, {}),
    function_timeout: Duration.minutes(2),
    function_security_groups: [],
    vpc: props.vpc,
    subnets_selection: props.vpc.selectSubnets({
      subnetType: ec2.SubnetType.PRIVATE_WITH_NAT,
    }),
  });
```

> **Lambdas**
>
> The CDKResourceInitializer custom component uses an AWS Lambda function. We will cover Lambda functions extensively later in this book when we get into serverless architecture. If you are unfamiliar with them, they are one-time-run codes in a container. Once the code is executed, the container dies. Lambda functions are the workhorses of serverless architecture. You can read more about them here: https://aws.amazon.com/lambda/.

Let's examine how we are instantiating the CDKResourceInitilizer construct:

1.   We create a CDKResourceInitilizer custom construct, which, in turn, spins up a Lambda function.

2.   We define the log retention and length of the Lambda function's execution timeout. Since we only have one data point to inject, we don't need anything longer than two minutes.

3.   The Lambda function is backed by a custom Dockerfile we are passing to the custom construct.

4.   We also pass the same VPC and subnet network details for our Lambda function to run.

Next, we define the RDS instance as a dependency for this CustomResource. This means RDS needs to be up and running before CustomResource is spun up:

```
initializer.custom_resource.node.addDependency(this.instance);
```

Follow the preceding command by allowing a connection from the Lambda function to the isolated subnet RDS is in:

```
this.instance.connections.allowFrom(
    initializer.function,
    ec2.Port.tcp(3306),
  );
```

Now, let's examine our `Dockerfile` for the trigger Lambda function located at `infrastructure/lib/constructs/RDS/init`:

```
FROM amazon/aws-lambda-nodejs:14
WORKDIR ${LAMBDA_TASK_ROOT}

COPY package.json ./
RUN npm install -only=production
COPY index.js ./
COPY script.sql ./

CMD [ "index.handler" ]
```

Here is what is happening:

1. We are basing the image on `https://hub.docker.com/r/amazon/aws-lambda-nodejs`. We need to do this since the Lambda runtime needs all the goodies in this image to function.

2. We are defining the working directory as `LAMBDA_TASK_ROOT`. Again, this is needed for the Lambda function to run.

3. Then, we are copying over the `package.json` and `index.js` files and running npm `install`.

4. `script.sql` is also copied over into the image so that the Lambda function can run it.

Let's also examine the handler code for the Lambda function located inside the `index.js` file:

```
try {
  const { config } = e.params;
  const { password, username, host } = await getSecretValue(
    config.credentials_secret_name,
  );
  const connection = mysql.createConnection({
    host,
    user: username,
    password,
    multipleStatements: true,
  });

  connection.connect();
```

The preceding code gets the connection parameters from AWS Secrets Manager (covered in the next section) and establishes a connection with the RDS instance.

Next, it reads the `script.sql` file from the working directory on the container and executes it:

```
const sqlScript = fs
    .readFileSync(path.join(__dirname, 'script.sql'))
    .toString();
const res = await query(connection, sqlScript);
```

If all is well, we return an OK response; if anything fails, an error response is returned:

```
return {
    status: 'OK',
    results: res,
};
} catch (err) {
return {
    status: 'ERROR',
    err,
    message: err.message,
};
}
```

If we go back to `infrastructure/lib/constructs/RDS/index.ts`, toward the end of the file, you will see the response of our Lambda function returned as a `CfnOutput` response once the stack is completed. This lets us check whether the data seeding step was successful:

```
/* ----------
 * Returns the initializer function response,
 * to check if the SQL was successful or not
 * ---------- */
new CfnOutput(scope, 'RdsInitFnResponse', {
    value: Token.asString(initializer.response),

});
```

The output is as follows:

```
Outputs:
Chapter4Stack.BackendURL = chapter4-lb-1113296104.us-east-1.elb.amazonaws.com
Chapter4Stack.FrontendURL = chapter-4-web-bucket-akemxdjqkl.s3.amazonaws.com
Chapter4Stack.RdsInitFnResponse = {"status":"OK","results":[{"fieldCount":0,"affectedRows":0,"insertI
d":0,"serverStatus":10,"warningCount":0,"message":"","protocol41":true,"changedRows":0},{"fieldCount"
:0,"affectedRows":0,"insertId":0,"serverStatus":10,"warningCount":0,"message":"","protocol41":true,"c
hangedRows":0},{"fieldCount":0,"affectedRows":1,"insertId":1,"serverStatus":2,"warningCount":0,"messa
ge":"","protocol41":true,"changedRows":0}]}
Stack ARN:
arn:aws:cloudformation:us-east-1:531551740863:stack/Chapter4Stack/1a0113e0-1c90-11ed-8ad1-12070fa947a
3

 +  Total time: 888.76s
```

Figure 4.9 – Examining the CDK outputs to see whether the database seeding was successful

Great – we got an OK status for the seeding job. Now, let's see how to store secret information such as database credentials in our CDK apps.

Storing secrets

As you might know, we have to create a set of credentials for our relational databases, such as MySQL, which we'll use in this case. You might be wondering, where do we store these secrets?

A secret is anything that you would want to avoid storing in your GitHub repository. It could be things such as the following:

- Usernames and passwords to external systems, databases, and so on
- Any cryptographic private keys
- Database table names, identifiers, hosts, and so on
- Anything else that would compromise the integrity of the system if exposed

The main point is that if someone gains access to your code, they would not be able to access the deployment and take over the application without also having provided relevant AWS credentials to deploy the software. The worst case scenario is if an attacker gets access to your Git repository is material damage occurring as a result of loss of intellectual property. You wouldn't want the attacker to gain access to information about every parameter of your customer's lives that is saved in databases. Saving credentials on Git has wrecked lives in the past. Don't fall victim to it.

In this project, we had to generate MySQL credentials. But how do we go about generating a password for MySQL, while not having it in the Git history? And how do we then share these secrets with relevant components securely?

Let's take a look at the `infrastructure/lib/constructs/RDS/index.ts` file again; you will find the following block of code:

```
const instance_id = 'my-sql-instance';
const credentials_secret_name = `chapter-4/rds/${instance_id}`;

this.credentials = new rds.DatabaseSecret(scope, 'MySQLCredentials', {
    secretName: credentials_secret_name,
    username: 'admin',
});
```

The `DatabaseSecret` construct of the `aws-rds` library receives a secret name and a username as parameters and creates an, erm, secret in the AWS `SecretsManager` service. In this case, the secret is a set of credentials with a username, as defined here, and a password that the `DatabaseSecret` construct automatically generates. There are plenty of configurations you can add to adjust the types of passwords generated, as well as secret rotation schedules and a lot more.

Now, we must assign this set of credentials to the RDS MySQL instance:

```
this.instance = new rds.DatabaseInstance(scope, 'MySQL-RDS-Instance',
{
    credentials: rds.Credentials.fromSecret(this.credentials),
```

> **Serverless**
>
> AWS SecretsManager is a serverless component, which means we don't worry about the underlying technology to store our secrets securely. The service rotates them regularly and manages their life cycle within stacks generated by our code. This is an immense leap forward in the software development paradigm. Resources that would have been spent on maintaining such systems in the past can now be utilized to write even more awesome IaC code with AWS CDK.

You can find out more about these configurations in the DatabaseSecret documentation at https://docs.aws.amazon.com/cdk/api/v1/docs/@aws-cdk_aws-rds. DatabaseSecret.html.

With our secret generated, we have to allow access to the various components in our code that need access to the MySQL database. One such component is the AWS Lambda function we used to seed the database right after it starts up. Here, we are passing down the secret name to CDKResourceInitilizer so that we know where to get the secret from:

```
const initializer = new CDKResourceInitializer(scope, 'MyRdsInit', {
    config: {
        credentials_secret_name,
    },
```

But doing this alone doesn't give the Lambda function access to the secret. It knows where to get it from, but it still doesn't have permission to read it. We have to explicitly give these permissions to the initializer function:

```
this.credentials.grantRead(initializer.function);
```

The ISecret interface's grantRead function is a helper function that assigns the IAM policy of secretsmanager:GetSecretValue to the target component. The same concept holds for the ECS application container. This time, we won't be using a helper function. Open the /infrastructure/ lib/chapter-4-stack.ts file; you will find the following block of code:

```
this.ecs.task_definition.taskRole.addToPrincipalPolicy(
    new PolicyStatement({
        actions: ['secretsmanager:GetSecretValue'],
        resources: [this.rds.credentials.secretArn],
    }),
);
```

That's it – both our Lambda function and ECS service can now read the database credentials from `SecretsManager` and perform their tasks!

ACM certificate

I owe you an apology. At the beginning of this chapter, we made you buy a domain and then validate the domain to be able to issue the ACM TLS certificate. The domain buying bit you had to do, but the ACM certificate verification part you didn't.

Curse me as you wish, but we wanted to show you how much time and effort you could save if you switched to AWS CDK. Let me show you what we mean. This whole drama is automatically achieved by the following single block of code, which can be found in the `infrastructure/lib/ACM/index.ts` file:

```
this.certificate = new Certificate(scope, 'Certificate', {
    domainName: domain_name,

    validation: CertificateValidation.fromDns(props.hosted_zone),
    subjectAlternativeNames: ['*.cdkbook.click'],
});
```

Yes, that is all it takes. CDK will do all the work needed in the background to validate the domain for ACM to issue the certificate and all the cleanup work after it's done. We're sorry and yes, you can go ahead and delete that manually generated certificate.

I sense a moment of awkward silence.

Glue code

To wrap up this chapter, let's look at how all of these constructs come together. We saw a portion of this when we covered secrets. Let's go ahead and open the `infrastructure/lib/chapter-4-stack.ts` file again:

```
this.route53 = new Route53(this, 'Route53');

this.acm = new ACM(this, 'ACM', {
  hosted_zone: this.route53.hosted_zone,
});
```

Here, we are referencing the hosted zone that was created alongside the domain you purchased, as well as initializing the ACM level 3 construct and passing the hosted zone as a parameter for it to do the rest of the magic.

Next, we are creating an AWS VPC to host the ECS application container and RDS database, and for the data-seeding Lambda:

```
this.vpc = new Vpc(this, 'MyVPC', {
    subnetConfiguration: [
      {
        cidrMask: 24,
        name: 'ingress',
        subnetType: SubnetType.PUBLIC,
      },
      {
        cidrMask: 24,
        name: 'compute',
        subnetType: SubnetType.PRIVATE_WITH_NAT,
      },
      {
        cidrMask: 28,
        name: 'rds',
        subnetType: SubnetType.PRIVATE_ISOLATED,
      },
    ],
  });
```

We have created an isolated private subnet to house the MySQL instance since isolated subnets don't have NAT gateways that route traffic to the internet, adding another layer of security. Subnets and their various types and configurations are outside the scope of this book. If you're not familiar with them, we recommend reading the official AWS docs (you can find the relevant page here: https://docs.aws.amazon.com/vpc/latest/userguide/configure-subnets.html) or you could look for some quick explainer videos on YouTube. We find the AWS docs way too comprehensive for beginners.

Now, we need some more boilerplate code to initialize S3, RDS, and ECS custom constructs:

```
this.s3 = new S3(this, 'S3', {
    acm: this.acm,
    route53: this.route53,
  });

  this.rds = new RDS(this, 'RDS', {
    vpc: this.vpc,
  });

  this.ecs = new ECS(this, 'ECS', {
    vpc: this.vpc,
    acm: this.acm,
    route53: this.route53,
  });
```

We also need to allow connections from the ECS container to the RDS database:

```
this.rds.instance.connections.allowFrom(this.ecs.cluster, Port.
tcp(3306));
```

Finally, we want to make sure the RDS database is up and running before kicking off the application container. So, we have to declare the RDS instance as a dependency. This can be achieved with this line of code:

```
this.ecs.node.addDependency(this.rds);
```

Summary

When writing this chapter, we were hoping to close the loops around a production-grade AWS CDK setup. In this chapter, we learned how to purchase domains in Route 53 and create hosted zones for them. We then went on a (necessary) wild goose chase validating a DNS to issue ACM certificates. We then realized how CDK saves us time by doing most of the work of DNS validation for us. We configured an RDS MySQL database for our TODO application, and we also learned about the importance of storing secrets safely and how to do that with CDK apps.

There are six more chapters in this book. That said, if you've gotten this far, you have the general knowledge to start creating awesome DevOps automations with AWS CDK. We hope the game-changing characteristics of CDK are becoming as evident to you as they are to us. In the next chapter, we will cover how to build bulletproof CI/CD pipelines for CDK applications. See you there!

5
Continuous Delivery with CDK-Powered Apps

You've surely heard of continuous **integration/continuous delivery** (**CI/CD**) and most likely practiced it yourself in your projects. Take the example from the last chapter. We developed and deployed a full stack web application using AWS CDK. But that's not where our work as developers stops. Software systems continuously evolve and need to incorporate changes to the code base. New features are added and bugs are fixed. New versions of software need to be rolled out.

The friction between developing code and deploying must be minimal, the reason being that having fewer deployments means that each deployment will have a lot of changes incorporated in it. With a lot of changes going on at once, you will have many ways in which things can go wrong. This ultimately causes a vicious circle, where more production releases go wrong and developers, as a result, become weary of releasing changes, which in turn exacerbates the problem of releases going wrong since each release will end up being packed with changes.

A principled and automated approach using CI/CD tools is the remedy to this situation. In this chapter, we are going to cover the following topics:

- An introduction to CI/CD and AWS's toolset
- Creating various environments for our application
- Using AWS's **CodeBuild** and **CodePipeline** to implement a robust CI/CD process
- Running the build for various branches
- Getting notifications of build status

Introduction to CI/CD

Before jumping into action, we wanted to take some time to introduce you to some concepts about CI/CD and explain why even though it's imperfect, AWS's CI/CD toolset fits CDK projects the best.

What is continuous integration (CI)?

Put simply, CI is the process of creating different branches for bug fixes and features in your projects. Once the results are satisfactory, you can go ahead and merge these changes into the main branch. By satisfactory results, we mean the code is reviewed and well tested.

While CI is in the most part a development procedure issue, there are elements of CI that need to be automated. We don't know about you, but we wouldn't be able to trust the word of every developer who has run the tests that everything was green and OK. This integration needs to happen automatically and the results must be reported.

What is continuous delivery (CD)?

Once the steps from the CI process are complete, CD kicks in by automatically deploying these changes to development, staging, and production environments. The changes are often propagated through the environments. They are *promoted* once they pass developer checks and the product side of the team has seen it in staging and given it the go-ahead. This promotion process varies from one team to the other. We ourselves am in the *least possible resistance to releases* camp but even for me, this changes from one project to another. A TODO application will have nowhere near the same level of scrutiny as a banking or public health software application.

AWS's CI/CD toolset

There are numerous fairly advanced CI/CD toolsets. Much like programming languages and frameworks, these tools come with a variety of features that might fulfill the needs of a particular kind of project. Some examples are GitHub Actions, Travis, and CircleCI.

What we will argue is that if CDK is your IaC framework of choice, you can stop shopping and stick with AWS's **CodeBuild** and **CodePipeline**. The reason? You guessed it: we can use CDK to define the CI/CD structure of our application. What's more, our CI/CD process can be easily integrated with all other AWS services using CDK. Want to send an email to administrators once a build is done? No problem. Want to kick off certain database migrations during the build? Or want to kick off a Lambda function that runs some sanity checks on the environments post deployment? Or one that tags your Git branch with release details? No problem, no problem, and no problem. While a lot of CI/CD tools out there bake these features in, AWS leaves all of that up to you. You are provided with a versatile toolset and the world is your oyster. We like this kind of freedom.

Now that we have an idea of what CI/CD is and how AWS's CodeBuild and CodePipeline differ from the rest of the offerings, let's get started and implement a simple yet robust CI/CD pipeline for our TODO application.

Technical requirements

Just like previous chapters, the source code for this chapter can be found in the following GitHub directory: `https://github.com/PacktPublishing/AWS-CDK-in-Practice/tree/main/chapter-5-continuous-integration-with-cdk-powered-apps`.

The Code in Action video for this chapter can be viewed at: `https://packt.link/vd91G`.

GitHub personal access token

CodePipeline is a CI/CD tool that needs to get the latest changes from the code repository. AWS has its own GitHub repository hosting platform named **CodeCommit**, which we would use in the blink of an eye had we not found the UI to be eye-wateringly horrendous (there goes our AWS partnership status), since like everything else, the integration with CodePipeline is a breeze.

For now, it's cool to stick to GitHub and we need to figure out a way for CodePipeline to access the GitHub repo and be informed of changes. Thankfully, CodePipeline does most of that for us with a **Source** step. This **Source** step needs a personal access token issued by GitHub to make the relevant API calls in the background and get hold of the repository changes.

To avoid triggering everyone's CI/CD pipeline whenever we push code to the main book repository, please go ahead and fork the chapter code into a private GitHub repository under your account. Remember to have only the `chapter 5` code in this forked repository.

Figure 5.1 – Forking the book's GitHub repository

On the next page, choose your personal profile (most likely selected by default) to fork the repository into, leave the name as is, and untick the **Copy the main branch only** box. Now comes the personal access token part:

1. Click on your GitHub profile picture and click **Settings**.

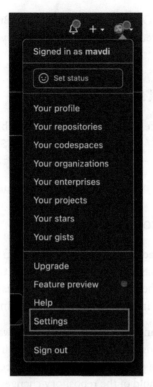

Figure 5.2 – Click on your profile picture and then Settings

2. In the menu on the left, click on **Developer settings**.

3. Next, click **Personal access tokens** and select **Fine-grained tokens**.

Figure 5.3 – Navigating the GitHub Settings page

4. At the top right, click the **Generate new token** button.

5. Fill in the token name, naming it `aws-cdk-book` or any other name to your liking.

6. Leave the expiry as **30 days**. It's best to remain on the safe side and expire tokens within a short time. If you need to come back to this chapter, you might have to generate another token depending on when you do it.

7. Under the **Repository access** section, pick **Only select repositories**. From the drop-down list, pick the repository you just forked. The name should be your GitHub profile name followed by **AWS-CDK-in-Practice**.

8. Under **Permissions**, click **Repository**. Here, we have to give CodeBuild the correct access so that it's able to be notified of code changes.

9. From the list, pick the following items and give CodeBuild read and write access: **Commit statuses**, **Metadata**, **Contents**, and **Webhooks**.

10. Click **Generate token**.

11. On the next page, you will receive a token starting with **github_pat_** followed by some random alphanumerical values. Copy it and keep it handy and safe.

Slack integration steps

To be notified of build statuses, we need a **Slack** workspace. If you currently don't have one, or can't change the configuration of your current workspace, please go ahead and create a workspace by signing up using the following URL: `https://slack.com/get-started#/createnew`.

Once the workspace is ready, we need to enable the **AWS Chatbot** Slack plugin. Here are the steps you need to take within the Slack app to do that:

1. Create a workspace or use one where you have the necessary administrative access to add an app to it.

2. Click on the name of the workspace on the left, go to **Settings & administration**, and then go to **Manage apps**.

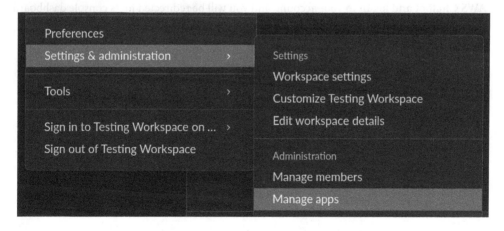

Figure 5.4 – Getting to the admin settings in Slack

3. You will be redirected to the Slack web page, where you can manage all of the workspace's apps. Once on this page, start by copying the URL and extracting the ID from it. This ID is your workspace ID so save it somewhere, such as on Notepad, as you will need it later for integrating with the pipeline notifications.

Figure 5.5 – Slack URL with the workspace ID

4. Next, on the same page, search for **AWS Chatbot**.

Figure 5.6 – Search for AWS Chatbot in Slack apps

5. You will be redirected to the app's page. Click **Add to Slack**.

6. You will then be redirected to the AWS Chatbot web page in AWS. Click on **Get started with AWS Chatbot and log in to** your account, and you will be redirected to its console dashboard.

7. For **Chat client**, choose **Slack** and click on **Configure client**. You will then be redirected to a Slack web page where you need to confirm this integration.

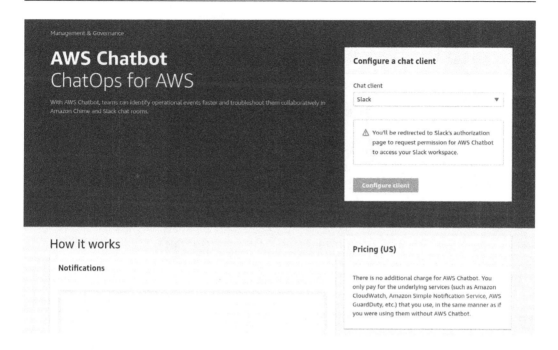

Figure 5.7 – Configuring Slack in the AWS dashboard

8. Once everything is finished, you will be redirected to the AWS dashboard again and, if you go back to the **Manage apps** page, you will now see a new app has been added to your Slack workspace.

9. Now, you can add AWS Chatbot to a channel. Create two channels, named **book-dev** and **book-prod**. Right-click on each channel and select **Copy | Copy linkink**. The channel ID for each channel can be found at the end of the copied URL, just like the workspace ID. Don't forget to save these values because we'll need them later.

Figure 5.8 – Copying the Slack channel link

10. Finally, to receive notifications in each channel, simply run the `/invite @AWS` command.

Figure 5.9 – Inviting AWS Chatbot in Slack channels

That's all it takes to enable AWS Chatbot for Slack. We're aware that this is a manual process, but you'll only have to do this once for all your projects within the same AWS account. Let's next look into creating different deployment environments for our app.

Creating various environments

In this chapter, for the sake of simplicity, we will be creating two different environments for our TODO application to be deployed in – **development** and **production**. You might have various other environments when working with software projects of your own, such as **staging** and **demo** environments. Additionally, you might also want to create a separate deployment for each pull request on GitHub, run a bunch of tests against it, and then spin down the deployment automatically. CI/CD pipelines can get very involved and we can't cover all the various ways you could build such pipelines, but the skills you learn in this chapter should cover most of what you need to know to get to advanced CI/CD implementations. Let's begin.

Open up the `chapter 5` code in your favorite code editor. At the root of the directory (so under `AWS-CDK-in-Practice-Chapter-5-directory`), create a file named `config.json`. Here is what mine looks like given that I purchased and configured the `cdkbook.click` domain name. Yours will be the domain you either purchased or configured in the last chapter. Keep the rest the same for now:

```
{
  "domain_name": "cdkbook.click", //change this to your domain name
  "backend_subdomain": "backend-cdk-book",
  "frontend_subdomain": "frontend-cdk-book",
  "backend_dev_subdomain": "dev-backend-cdk-book",
  "frontend_dev_subdomain": "dev-frontend-cdk-book"
}
```

Next, under the `infrastructure` directory, create two files named `.env.development` and `.env.production` (don't forget the dots at the beginning).

For `.env.development`, copy in the following text, with `GITHUB_TOKEN` being the personal access token you generated in the previous section:

```
GITHUB_TOKEN=<YOUR_PERSONAL_ACCESS_TOKEN>
CHANNEL_ID=<BOOK_DEV_CHANNEL_ID>
WORKSPACE_ID=<YOU_SLACK_CHANNEL_ID>
NODE_ENV=Development
```

Do the same for the `.env.production` file, except the `NODE_ENV` property for this file will be set to `Production`:

```
GITHUB_TOKEN==<YOUR_PERSONAL_ACCESS_TOKEN>
CHANNEL_ID=<BOOK_PROD_CHANNEL_ID>
WORKSPACE_ID=<YOU_SLACK_CHANNEL_ID>
NODE_ENV=Production
```

We've created these files because we are using the `dotenv` library (found at `https://www.npmjs.com/package/dotenv`). This package loads the environment variables for an application execution through `.env` files, one of which we have created for each of the **Development** and **Production** environments we intend to deploy to. We've not made extensive use of it except for using the personal access token and the `NODE_ENV` variable for naming CDK objects. In larger projects, `.env` files are a lot more useful. There could be cases where you would have an initial script in your CI steps that reads some secrets and tokens from **AWS Secrets Manager** and saves them in a `.env` file to be later picked up by the CI/CD pipeline or the application itself.

Implementing a robust CI/CD process

We just have one more thing to do before we can start the pipeline stack. We need to update the pipeline source to use the forked repo. Go to `infrastructure/lib/constructs/Pipeline/index.ts` and then scroll down to line 190, where you will find this code:

```
this.pipeline.addStage({
  stageName: 'Source',
  actions: [
    new GitHubSourceAction({
      actionName: 'Source',
      owner: <your_github_user>,
      repo: <the_repository_name>,
      branch: `${branch}`,
      oauthToken: secretToken,
      output: outputSource,
      trigger: GitHubTrigger.WEBHOOK,
    }),
  ],
});
```

Make sure to update the values for the `owner` and `repo` properties to match your own before deploying the pipeline. This ensures that CodePipeline will search for the correct GitHub repository. We will talk more about the **Source** stage later in the chapter.

Now that we've done all the prepping, it's time for the magical moment. Open up the `infrastructure` directory in the terminal shell and install `yarn` packages using just the following command:

```
$ yarn
```

After this, simply deploy the CI/CD pipeline using the following command:

```
$ yarn run cdk:pipeline deploy --profile cdk
```

That's right, we have defined the entire CI/CD pipeline using AWS CDK. When prompted, enter *Y* and press *Enter*. Let CDK do its magic and create the pipeline for you. While that happens, let's open up `package.json` under the `infrastructure` directory and see what this `yarn` command does. Look within the `scripts` section and run the following command:

```
"cdk:pipeline": " "cross-env CDK_MODE=ONLY_PIPELINE cdk"
```

When you type in the `yarn run cdk:pipeline` command, `yarn` then translates it into the preceding command. That command calls the `cdk` CLI with a `CDK_MODE=ONLY_PIPELINE` environment variable.

If you now look into `infrastructure/bin/chapter-5.ts`, you will see that this environment variable gets checked and if it's set, only the pipeline CDK **stack** will be deployed:

```
if (['ONLY_PIPELINE'].includes(process.env.CDK_MODE || '')) {
  new Chapter5PipelineStack(app, 'Chapter5PipelineStack', {
    env: {
      region: process.env.CDK_DEFAULT_REGION,
      account: process.env.CDK_DEFAULT_ACCOUNT
    },
  });
}
```

"Great, so how do we then deploy the main stacks?" you ask. Well, the pipeline does that for you! To see this, log on to your AWS console and find **CodePipeline** in the list of services. There you should be seeing two pipelines created, named `Chapter5-Development` and `Chapter5-Production`, which will most likely have a status of `In Progress` by the time you get there. Wait for them to change status to `Succeeded`:

Figure 5.10 – Successfully run pipelines

Now, let's test out the application. Pick the domain name you added to the configuration and add `dev-frontend-cdk-book` to it; in our case, that is `dev-frontend-cdk-book.cdkbook.click`:

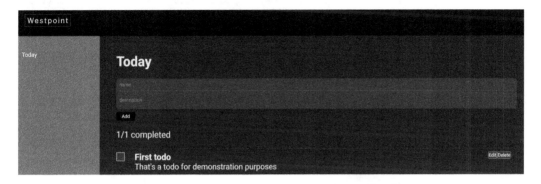

Figure 5.11 – Successfully loading the development environment's frontend

That's your **Development** environment! Now, remove the `dev-` bit from that URL and you will get the **Production** environment, which in this case should look exactly like the **Development** environment. But they are separate environments. You can certify this by adding a TODO item to the list in one of the environments. It won't appear on the other. Now, let's go into **CloudFormation** from the list of AWS services and see what is happening there:

Stack name	Status	Created time	▼
● Chapter5Stack-Production	⊘ CREATE_COMPLETE	2022-10-25 11:17:16 UTC-0300	
● Chapter5PipelineStack	⊘ CREATE_COMPLETE	2022-10-25 11:07:17 UTC-0300	
● Chapter5Stack-Development	⊘ CREATE_COMPLETE	2022-10-23 17:51:57 UTC-0300	

Figure 5.12 – List of stacks in CloudFormation

We can see in the list of stacks that as well as **Chapter5PipelineStack**, which is what we deployed, there are two other stacks that are created by the pipeline stack for each of the **Development** and **Production** environments. Essentially, our pipeline stack gives birth to and then maintains the two other stacks!

Let's go back to CodePipeline and examine one of the pipelines created. We'll go with `Chapter5-Development`. There you will see four stages, hopefully all green, named **Source**, **Back-End-Test**, **Front-End-Test**, and **Build-and-Deploy**.

> **If you're facing an issue**
>
> If you run into any unexpected behavior, don't hesitate to let us know by raising an issue on GitHub. You can also reach out to us using the contact details provided in the preface of the book.

Here is a brief explanation of what each stage does:

- **Source**: This is the step that is responsible for reading the repository and having the rest of the pipeline triggered when the GitHub repository is updated. This is also the step that uses the personal access token that you generated.

- **Back-End-Test**: This step installs the `yarn` dependencies of the backend, or the `server` directory, and subsequently runs the backend tests. If any of those tests fail, this step fails, and the pipeline is aborted.

- **Front-End-Test**: This does a job very similar to that of the previous step except for the frontend, or the `web` directory. Again, should any of these tests fail, the pipeline stops.

- **Build-and-Deploy**: Pretty much does what the name says. We will get into the details but it pretty much does what you used to do to deploy the main stacks: it runs the `cdk deploy` command.

Let's now examine how all this comes to happen. Let's start by looking at the main `infrastructure` file found under `bin/chapter5.ts`:

```
const app = new cdk.App();

if (['ONLY_DEV'].includes(process.env.CDK_MODE || '')) {
  new Chapter5Stack(app, `Chapter5Stack-${process.env.NODE_ENV ||
''}`, {

    env: {
      region: process.env.CDK_DEFAULT_REGION,
      account: process.env.CDK_DEFAULT_ACCOUNT,
    },
  });
}

if (['ONLY_PROD'].includes(process.env.CDK_MODE || '')) {
  new Chapter5Stack(app, `Chapter5Stack-${process.env.NODE_ENV ||
''}`, {

    env: {
      region: process.env.CDK_DEFAULT_REGION,
      account: process.env.CDK_DEFAULT_ACCOUNT,
    },
  });
}

if (['ONLY_PIPELINE'].includes(process.env.CDK_MODE || '')) {
  new Chapter5PipelineStack(app, 'Chapter5PipelineStack', {

    env: {
```

```
        region: process.env.CDK_DEFAULT_REGION,
        account: process.env.CDK_DEFAULT_ACCOUNT,
      },
    });
  }
```

Here, we see that there are three main stacks defined: the almighty `Chapter5PipelineStack`, and the two other main stacks for the **Development** and **Production** environments. Each of these stacks is only spun up if the CDK_MODE environment values are either ONLY_DEV, ONLY_PROD, or ONLY_PIPELINE.

This is why, when we defined the `cdk:pipeline yarn` command, we defined it as follows:

```
"cdk:pipeline": " "cross-env CDK_MODE=ONLY_PIPELINE cdk"
```

This sets the CDK_MODE environment variable to ONLY_PIPELINE and then leaves the building of the two other stacks to the pipeline stack.

> **cross-env**
>
> If you are not familiar with `cross-env`, it is a JavaScript library that lets you define environment variables before running a command for different platforms, including Windows, macOS, and Linux.

So, essentially, we have a mechanism that is explained by the following graphic:

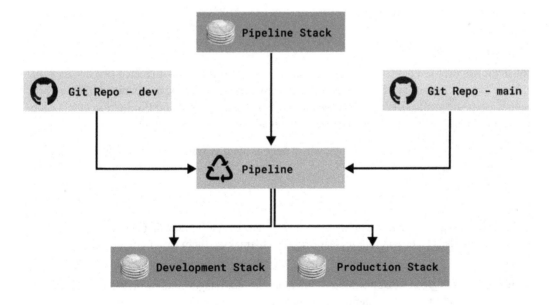

Figure 5.13 – How the stacks are generated

Looking at the preceding diagram, here is what is happening:

1. We manually deploy the pipeline stack, which in turn creates the pipelines.

2. Pipelines are then triggered when a branch is updated or at the beginning of the pipeline creation.

3. These pipelines in turn create the **Development** and **Production** environments based on the dev and main git branches respectively.

As we saw, once the CDK_MODE environment variable is set to ONLY_PIPELINE, then Chapter5PipelineStack gets initiated. Let's look at what happens within this stack. Open the infrastructure/lib/chapter-5-pipeline-stack.ts file:

```
export class Chapter5PipelineStack extends Stack {
  constructor(scope: Construct, id: string, props: PipelineProps) {
    super(scope, id, props);

    /* ---------- Constructs ---------- */
    new PipelineStack(this, 'Chapter5-Pipeline-Prod', {
      environment: 'Production',
    });

    new PipelineStack(this, 'Chapter5-Pipeline-Dev', {
      environment: 'Development',
    });

    /* ---------- Tags ---------- */
    Tags.of(scope).add('Project', 'Chapter5-Pipeline');
  }
}
```

After all the imports and bootstrapping are done, we see that this stack has created two instances of the PipelineStack class with the passing of the environment variable as a parameter. These two pipelines are nearly identical, except for a few places where **Development** and **Production** differ in the way the project is built and deployed. Hence, it makes sense that we derive both from the same class.

Finally, at the end we are adding CDK tags, which help us identify resources, especially when dealing with billing and charges.

> **Tags**
> We will cover tags in more detail in the next chapter. For now, just know that with the hundreds of resources that a CDK app might create, tags make tracking them easier.

Now, let's see what PipelineStack does. Remember the build stages you saw in AWS CodePipeline? Well, unsurprisingly, the code to define them looks pretty similar. Open up infrastructure/

`lib/Pipeline/index.ts` and look toward the end of the file where we define the stages. Let's go through them one by one.

This is where we define the **Source** stage:

```
this.pipeline.addStage({
    stageName: 'Source',
    actions: [
      new GitHubSourceAction({
        actionName: 'Source',
        owner: '<change to your username>',
        repo: '<change to your repo name>',
        branch: `${branch}`,
        oauthToken: secretToken,
        output: outputSource,
        trigger: GitHubTrigger.WEBHOOK,
      }),
    ],
});
```

This stage is mostly self-explanatory, where we first locate the repository and the branch in question by passing down the details for the `owner`, `repo`, and `branch` properties. The `secretToken` property is defined at the top as follows:

```
const secretToken = new SecretValue(process.env.GITHUB_TOKEN);
```

We are basically creating a Secrets Manager key pair by reading the `GITHUB_TOKEN` environment variable and storing it in there. This secret can then be safely accessed by our pipeline given the right permissions. Next, let's take a look at the constructor of the `PipelineStack` class:

```
constructor(scope: Construct, id: string, props: Props) {
    super(scope, id);
    const {
      path,
      buildCommand,
      deployCommand,
      branch,
      tag,
    } = pipelineConfig(props.environment);
```

The `branch` variable is a value passed down through a `pipelineConfig` utility function found at `infrastructure/lib/utils/pipelineConfig.ts`, which grabs the values in the `config.json` file you created in the root of the project.

Depending on the environment value passed down through the constructor, it adds some environment-specific configurations with `branch` being one of the values. Since the building and deploying commands will be different for each environment, these values are also passed down from the `pipelineConfig` utility function. The following shows what that utility function looks like:

```
export const pipelineConfig = (env: string) => {
  if (env === 'Production') {
    return {
      path: '.env.production',
      buildCommand: 'yarn build:prod',
      deployCommand: 'yarn cdk deploy',
      branch: 'master',
      tag: 'chapter5-production-pipeline',
    };
  }

  return {
    path: '.env.development',
    buildCommand: 'yarn build:dev',
    deployCommand: 'yarn cdk:dev deploy',
    branch: 'dev',
    tag: 'chapter5-development-pipeline',
  };
};
```

> **Code structuring**
>
> The code in this book is structured in a way that makes it easier for TypeScript beginners to understand and follow, but that by no means is the optimal way to structure your CDK code. There are interesting patterns forming in the CDK community, an example of which is dependency injection using libraries such as **TSyringe** (see `https://github.com/microsoft/tsyringe`).

The `output` property of the `Source` stage basically hands over the contents of the repository to the next stage so that we can start running the next steps. Speaking of the next stage, here is how we configure the **Back-End-Test** stage:

```
this.pipeline.addStage({
    stageName: 'Back-End-Test',
    actions: [
      new CodeBuildAction({
        actionName: 'Back-End-Test',
        project: this.backEndTestProject,
        input: outputSource,
        outputs: undefined,
      }),
    ],
  });
```

This stage is mostly self-explanatory, with the input being what the previous stage set, but more importantly, there is a `this.backendTestProject` variable that it reads. Let's see what this variable entails:

```
this.backEndTestProject = new PipelineProject(
      scope,
      `Chapter5-BackEndTest-PipelineProject-${props.environment}`,
      {
        projectName: `Chapter5-BackEndTest-PipelineProject-${props.
environment}`,
        environment: {
          buildImage: LinuxBuildImage.fromCodeBuildImageId(
            'aws/codebuild/amazonlinux2-x86_64-standard:4.0',
          ),
        },
        buildSpec: BuildSpec.fromObject({
          version: '0.2',
          phases: {
            install: {
              'runtime-versions': {
                nodejs: '16',
              },
            },
            pre_build: {
              'on-failure': 'ABORT',
              commands: ['cd server/', 'yarn install'],
            },
            build: {
              'on-failure': 'ABORT',
              commands: ['echo Testing the Back-End...', 'yarn test'],
            },
          },
        }),
      },
    );
```

This code created a CodePipeline project where we spin up a build machine that runs on an EC2 machine through a container, with given commands from us to run within the code project that the **Back-End-Test** step imported. Sounds complex, right? The good news is that you won't have to deal with the complexities of running and shutting down the build machine. CodePipeline does all that for you. Let's go through all the properties of this pipeline project:

- The `projectName` property is composed of the name of the chapter and the `{props.environment}` variable, which is passed down through the constructor.

- The `environment` property describes the machine. Here we are creating an `aws/codebuild/amazonlinux2-x86_64-standard:4.0`-based image. There are a lot more configurations you can do here. Take a look at the docs found at `https://docs.aws.amazon.com/cdk/api/v1/docs/@aws-cdk_aws-codebuild.BuildSpec.html`.

- The `buildSpec` property encapsulates what you can do through AWS CodePipeline buildSpec files into an object. There are lots of options to configure here (see `https://docs.aws.amazon.com/codebuild/latest/userguide/build-spec-ref.html`) but here are the ones we need to configure in this example:

  - `install` is a step that installs the tooling we need to run the tests, in this case, Node.js runtime version 16.

  - `pre_build` is a step for developers to add their own pre-build steps. In our case, we are running `yarn install` within the `server` directory.

  - `build` is where we define the actual step to run the test via `yarn test`. In both this step and the previous one, we instruct CodePipeline to abort the build in case of failures, meaning no deployment of software if our tests fail!

> **Documentation**
>
> Don't forget to look into AWS CDK's official documentation to find out about various ways to configure constructs (found at `https://docs.aws.amazon.com/cdk/api/v2/`). You can also get a good idea of the input properties by pressing *Ctrl + Click* (Mac: *Command + Click*) on the property names in **VSCode** and many other code editors.

The **Front-End-Test** stage is also built in a similar fashion, with the `pre_build` and `build` steps running within the web directory:

```
pre_build: {
    'on-failure': 'ABORT',
    commands: ['cd web/', 'yarn install'],
    },
    build: {
        'on-failure': 'ABORT',
        commands: ['echo Testing the Front-End...', 'yarn test'],
    },
```

The final stage is **Build-and-Deploy**, which runs the `deployProject` CodeBuild instructions. These are the most complicated of the build steps, but they are still more or less the same steps you would run locally to build and deploy the CDK project. Let's examine the steps. Here we are dividing the stage into four phases, the first one being `install`:

```
install: {
    'runtime-versions': {
```

```
        nodejs: '16',
    },
  },
```

The `install` build stage defined here is the same as the `pre_build` phase of the previous build projects. In the `pre_build` phase here we are running `cd` in the web, server, and infrastructure phases, and running `yarn install` within them to have the Node.js packages of the project ready. Additionally, we are setting certain environment variables the frontend needs to function alongside the backend:

```
pre_build: {
    'on-failure': 'ABORT',
    commands: [
      'cd web',
      'yarn install',
      `

      echo '{
        "domain_name": "${domainName}",
        "backend_subdomain": "${backendSubdomain}",
        "frontend_subdomain": "${frontendSubdomain}",
        "backend_dev_subdomain": "${backendDevSubdomain}",
        "frontend_dev_subdomain": "${frontendDevSubdomain}"
      }' > src/config.json
      `,
      'cd ../server',
      'yarn install',
      'cd ../infrastructure',
      'yarn install',
    ],
  }
```

These environment variables are again defined within the `pipelineConfig.ts` file we looked at earlier. The same configuration for `backendSubdomain` is used when creating the `ARecord` for the backend within the `infrastructure/lib/ECS/index.ts` file:

```
const backEndSubDomain =
    process.env.NODE_ENV === 'Production'
      ? config.backend_subdomain
      : config.backend_dev_subdomain;

  new ARecord(this, 'BackendAliasRecord', {
    zone: props.route53.hosted_zone,
    target: RecordTarget.fromAlias(
      new LoadBalancerTarget(this.load_balancer),
    ),
```

```
        recordName: `${backEndSubDomain}.${config.domain_name}`,
    });
```

In this way, the backend will know how to create the `ARecord` for the API and the frontend will expect the same DNS route when making the API calls.

Continuing with the build project, depending on the environment we are in, the build command will differ, also defined within the `pipelineConfig.ts` file:

```
buildCommand: 'yarn build:prod',
deployCommand: 'yarn cdk deploy',
...
buildCommand: 'yarn build:dev',
deployCommand: 'yarn cdk:dev deploy',
```

These commands are in turn used in this build phase:

```
build: {
    'on-failure': 'ABORT',
    commands: [
      'cd ../web',
      `${buildCommand}`,
      'cd ../infrastructure',
      `${deployCommand}`,
    ],
}
```

To make this all function, CodeBuild needs to be able to be given the right roles to deploy stuff on CloudFormation, within the same `pipeline/index.ts` file where you will find these permissions:

```
const codeBuildPolicy = new PolicyStatement({
      sid: 'AssumeRole',
      effect: Effect.ALLOW,
      actions: ['sts:AssumeRole', 'iam:PassRole'],
      resources: [
        'arn:aws:iam::*:role/cdk-readOnlyRole',
        'arn:aws:iam::*:role/cdk-hnb659fds-lookup-role-*',
        'arn:aws:iam::*:role/cdk-hnb659fds-deploy-role-*',
        'arn:aws:iam::*:role/cdk-hnb659fds-file-publishing-*',
        'arn:aws:iam::*:role/cdk-hnb659fds-image-publishing-role-*',
      ],
    });
...
// adding the necessary permissions in order to synthesize and deploy
the cdk code.
this.deployProject.addToRolePolicy(codeBuildPolicy);
```

The roles we've assigned are preconfigured AWS roles. The name has some randomness to it, but here is the documentation link pointing to these roles: `https://docs.aws.amazon.com/cdk/v2/guide/bootstrapping.html`.

The rest of the code deals with integrating Slack to enable a full notification system for the pipeline. Whenever there's a change in the pipeline's state, a notification will be sent to the channels you selected.

Let's go through its code. Scroll down until the end of the `infrastructure/lib/constructs/Pipeline/index.ts` file and you will see this code:

```
const slackConfig = new SlackChannelConfiguration(this,
'SlackChannel', {
  slackChannelConfigurationName: `${props.environment}-Pipeline-Slack-
Channel-Config`,
  slackWorkspaceId: workspaceId || '',
  slackChannelId: channelId || '',
});
```

In this part, we are using an AWS Chatbot L3 construct named `SlackChannelConfiguration` to specify which Slack workspace and channel we want to link the pipeline with.

You will notice this segment of code that handles all the required configurations for sending notifications:

```
const snsTopic = new Topic(
this,
`${props.environment}-Pipeline-SlackNotificationsTopic`,
);
const rule = new NotificationRule(this, 'NotificationRule', {
  source: this.pipeline,
  events: [
    'codepipeline-pipeline-pipeline-execution-failed',
    'codepipeline-pipeline-pipeline-execution-canceled',
    'codepipeline-pipeline-pipeline-execution-started',
    'codepipeline-pipeline-pipeline-execution-resumed',
    'codepipeline-pipeline-pipeline-execution-succeeded',
    'codepipeline-pipeline-manual-approval-needed',
  ],
  targets: [snsTopic],
});


rule.addTarget(slackConfig);
```

Here, we're setting up an SNS topic to handle all the pipeline events that we want to send to Slack. We're also adding the previously configured Slack integration as the target for this topic. If you'd like to learn more about SNS, you can check out this link: `https://aws.amazon.com/sns/`.

That's it – we have a robust build system working. Now, every time you push your GitHub repo to `master` or `dev` branches, these build pipelines will run. As always, have your fun and when you're done with everything, make sure you kill both the pipeline and the main stacks by running this command within the `infrastructure` directory:

```
$  cdk destroy --profile cdk
```

After you're done, check CloudFormation in the AWS Management Console and make sure the pipeline, frontend, and backend stacks are all destroyed, unless you feel like giving AWS free cash. There is also more to explore within the code of this chapter. Both the `web` and `server` directories have tests declared this time, which are run by the `yarn test` command within each directory. Spend some time if you can and get familiar with the top-down structure of how everything is wired together.

Summary

This has been one of the most important chapters in this book from the end-to-end CDK project perspective. We learned how to set up GitHub and CodePipeline to work together. We created a root `Chapter5PipelineStack` that pulls code from GitHub, verifies the integrity of the application by running tests, builds both the frontend and backend portions, and deploys the stacks in multiple environments. As we said earlier in the chapter, your pipeline setup will probably be a lot more advanced than this in your production projects. Yet, the knowledge you've gained in this chapter should equip you to deal with these complexities and build awesomely automated infrastructures.

In the next chapter, we will look into tightening our CDK infrastructure with actual infrastructure tests! We will also review some of the practical lessons our teams have had to learn when dealing with CDK projects. See you there!

6

Testing and Troubleshooting AWS CDK Applications

Testing is a crucial part of building reliable and performant applications, and developing CDK applications is no exception to that. If you speak to the developers working with me (Mark), they will make jokes about my obsession with writing tests. We don't even necessarily enjoy the process of writing tests, although we must admit that watching green ticks appear when test scenarios are run is fun.

In this chapter, we will dive into the world of testing in the context of AWS CDK applications. Automated testing is essential for running a sanity check on your application and ensuring that the functionality and state of your application are as expected, even when new components are added or changed. We will focus on automated testing since manual testing can be time-consuming and error-prone, especially as CDK apps become more complex.

The primary purpose of automated testing is to run through a checklist of tests as quickly as possible, allowing changes to be incorporated quickly and software to evolve. As the complexity of your CDK app's logic increases, it becomes more critical to test your code automatically. Depending on your cloud architecture, your stack might be highly sensitive to certain changes. With multiple programmers working on a project, it is crucial to ensure that code and stack changes do not adversely affect your CDK infrastructure.

By the end of this chapter, you will have a good understanding of how to test your CDK applications automatically, ensuring that your code is performing as expected and preventing adverse impacts on your infrastructure. We will cover different types of tests, strategies for writing effective tests, and best practices for integrating testing into your CDK development workflow. We will also cover other means of troubleshooting your application by examining logs and debugging the code.

In this chapter, we will learn about the following topics:

- Various ways of testing CDK apps

- Examining CDK deployment logs

- Storing application logs

- Debugging CDK apps using VSCode

- Troubleshooting common problems

Technical requirements

Same as the previous chapters, you can find the code for this chapter in this book's GitHub repository at `https://github.com/PacktPublishing/AWS-CDK-in-Practice/tree/main/ chapter-6-testing-and-troubleshooting-cdk-applications`.

The code in this chapter is identical to the code in *Chapter 5* except for the addition of some tests, which we will go through in this chapter.

The Code in Action video for this chapter can be viewed at: `https://packt.link/Tp5R8`.

Understanding the terms of testing

First, let's get some of the definitions out of the way before we continue. Let's start with the types of tests:

- **Unit tests:** Unit tests involve testing every function individually and making sure that, given a certain expected and unexpected input, the function responds with an expected output or a graceful error.

- **Integration tests:** Integration tests are a higher level of testing than unit tests and often involve various modules for each scenario. Some integration testing scenarios might involve various functions and modules that make sure these functions work well in tandem to deliver certain functionality.

- **End-to-end (E2E) tests:** These tests provide the highest level of testing, which often involves running test scenarios against a live system.

- **Regression testing:** These types of tests are run to find inconsistencies and broken behavior when the system changes.

I won't go about arguing the case for and against any type of testing. What we've found over the years while working on various types of software products is that the higher the level of testing, the more useful it is to find issues in the code base. This makes sense since E2E testing involves many components, functions, and modules, all of which must be in the correct form for a scenario to successfully pass. That said, you might find cases where an E2E scenario passes but the system still malfunctions when it's given a different input. Often, in the projects that we work on, we will combine E2E and regression testing. Our E2E tests become a way for us to find regression issues. This is not limited to E2E testing – unit and integration tests can also detect regression issues.

For me, the value of writing tests can be summarized into these two points:

- Tests are a checklist of previous functionality. When they pass, we know that the new functionality that's been added hasn't broken the previous features.

- With added confidence in the code base, developer productivity increases. We no longer worry as much about inadvertently introducing bugs to the code base.

That said, we should be careful about wasting time on tests that don't add much value to team productivity. We've often seen teams of developers dedicate too much of their time to unit testing every single aspect of the code base. It's nice to add these checks but there will be trouble if a team spends more time writing and maintaining test scenarios as opposed to delivering features.

Moreover, there are various types of tests and multiple ways of integrating testing into development workflows. You might have heard of or practiced **test-driven development** (**TDD**) and **behavior-driven development** (**BDD**). The pros and cons of these methods are outside the scope of this book and are often affected by what the development team wants to do.

Various ways of testing CDK apps

CDK testing is generally performed in one of two ways:

- **Fine-grained assertions**: These types of tests are similar to unit testing and are performed to detect regressions. They are conducted with the AWS CDK `assertions` module (found at `https://docs.aws.amazon.com/cdk/api/v2/docs/aws-cdk-lib.assertions-readme.html`), which integrates with common testing frameworks such as Jest (`https://jestjs.io/`) and essentially tests that certain modules in our CDK app have certain properties – for example, a certain EC2 instance will have a specific AMI.

- **Snapshot tests**: Closer to integration testing, snapshot tests don't care how you wire up your CDK application. They compare the output of a synthesized CDK stack to a previous state after changes have been applied. Snapshot tests are a great way to refactor your CDK application while making sure the outputs are the same (another way to deal with regression issues).

AWS recommends using a combination of these tests to ensure the integrity of a CDK application. Snapshot tests are very interesting since they leave us free to refactor our code as we want, just being interested in the result of the synthetization process. The only headache with these tests is that the synthetization snapshots might also be different when a CDK library upgrade also happens. That said, if the library upgrades are done in isolation from other stack changes, snapshot tests are a great way to quality-assure CDK apps.

Now, let's take a quick look at how these tests are performed.

Running the tests

This chapter's code comes with a few fine-grained tests and snapshot tests preconfigured so that you can go through the exercise of running these tests. These tests are by no means exhaustive (meaning covering all the functionality) and this might not be the optimal way to configure tests in your project. However, hopefully, this will inspire you to improve. Let's take a look at how to run this chapter's tests.

Configuring the environments

Just like in the previous chapter, when we were running our CDK stack for various environments, we need to create a .env file in the infrastructure directory of chapter 6 to pass some values about the deployment to test the stack. We will name this file .env.testing so that it only gets picked up when we are running the tests. So, go ahead and create a .env.testing file under the infrastructure directory and fill it with values similar to the ones in the *Chapter 5* .env files:

```
CDK_DEFAULT_ACCOUNT=<Your 12 Digit AWS Account ID>
CDK_DEFAULT_REGION=us-east-1
```

CDK needs these environment variables to fill up certain values in the CloudFormation output. These same values could be fed into the stack with a --profile flag when deploying the stack, but since we are running tests using the jest and yarn commands here, we need to pass them in using environment variables.

> **AWS account IDs**
>
> Your AWS account ID is a 12-digit unique identifier. The steps to retrieve the value from the AWS console or CLI tooling can be found at https://docs.aws.amazon.com/accounts/latest/reference/manage-acct-identifiers.html.

Next, we need to build the frontend code. To do that, cd to the web folder and run the following command:

```
$ yarn
```

Follow this up with the following command:

```
$ yarn build:dev
```

Next, under the infrastructure directory, with all the dependencies installed, just run the following command:

```
$ yarn test
```

You should be able to see an output informing you that a total of three assertion tests and one snapshot were successfully run:

```
mavdi@Ms-MacBook-Pro-3 infrastructure % yarn test
yarn run v1.22.19
$ yarn build && jest
$ tsc
 PASS   test/chapter-6-stack.test.js
  Testing Chapter 6 code.
    ✓ The stack has a ECS cluster configured in the right way. (155 ms)
    ✓ The stack has a RDS instance configured in the right way. (85 ms)
    ✓ The stack has the VPC configured in the right way. (82 ms)
    ✓ Matches the snapshot. (84 ms)

Test Suites: 1 passed, 1 total
Tests:       4 passed, 4 total
Snapshots:   1 passed, 1 total
Time:        1.157 s, estimated 2 s
Ran all test suites.
✨  Done in 5.44s.
```

Figure 6.1 – Running the test suite and seeing the resulting passing tests

The yarn test command (which any JavaScript/TypeScript developer is most likely already familiar with) is hooked to the test script definition in package.json, which in our case runs the following:

```
$ yarn build && jest
```

Here, yarn build runs the build process of the project. The && operator in Bash is used to execute a command but only if the preceding command succeeded. In this case, this line means that if the project was successfully built, we will run the jest command since Jest is configured to be our testing framework. This configuration is incredibly easy. Jest will look for any filename ending in .test.ts, provide the test script with its API to run functions such as describe and test, and then run the scenarios accordingly.

Let's look at how these tests are configured. As mentioned earlier, this is the same CDK application from *Chapter 5*, with additional tests that you will find in the infrastructure/test directory. You will see a file named chapter-6-stack.test.ts, which contains our testing logic, and a directory named __snapshots__, which is autogenerated by CDK's tooling when the snapshot tests are initially run. Let's examine the test file first:

```
describe('Testing Chapter 6 code.', () => {
  // Using assertion tests:
  test('The stack has a ECS cluster configured in the right way.', ()
=> {
    // some testing code
  });

  test('The stack has a RDS instance configured in the right way.', ()
=> {
```

```
    // some testing code
  });

  test('The stack has the VPC configured in the right way.', () => {
    // some testing code
  });

  // Using snapshot tests:
  it('Matches the snapshot.', () => {
    // some testing code
  });
});
```

The first three tests are assertion tests that we declared using `test`, while the last test (the snapshot test) is declared using the `it` function. Both of these functions work the same way, but we used them here to distinguish the two tests. Let's cover the assertion tests first.

Assertion tests

Assertion tests are a type of validation that confirms that a piece of code is performing as intended. In the context of AWS CDK, these tests can be used to ensure that the resources being manipulated by CDK constructs are being created, modified, or deleted correctly. Other potential uses for assertion tests in AWS CDK include verifying the correct usage of AWS services, the correct setting for permissions, and the proper deployment of CDK apps and stacks. These tests involve defining a predetermined set of expected properties for the resources that a CDK stack will create and then comparing them to the actual properties of the created resources. The test passes if the actual properties match the expected ones and fails if they do not match. In the case of a failed test, the developer is alerted to investigate and resolve the issue.

Let's examine the first test in the preceding block. This test aims to make sure the ECS cluster we've created for our TODO application is configured in the way we intended it to be. To do this, we must set up the test by creating a CDK App and an instance of `Chapter6Stack`:

```
const app = new App();

  const chapter6Stack = new Chapter6Stack(app, 'Chapter6Stack', {
    env: {
      region: parsed?.CDK_DEFAULT_REGION,
      account: parsed?.CDK_DEFAULT_ACCOUNT,
    },
  });
```

The first line in the following block is a way for us to extract the CloudFormation template from the CDK stack so that we can run certain assertions against it:

```
const template = Template.fromStack(chapter6Stack);

template.resourceCountIs('AWS::ECS::Cluster', 1);

template.resourceCountIs('AWS::ECS::TaskDefinition', 1);

template.resourceCountIs('AWS::ECS::Service', 1);
```

We've just created some simple assertions. Here, we are analyzing the template to make sure a single instance of an **ECS Cluster**, an **ECS TaskDefinition**, and an **ECS Service** are created.

Our next assertion goes deeper than this. The template.hasResourceProperties function makes sure that a certain resource – in this case, a TaskDefinition class – has the properties we define in our assertion. Some of these properties can include the memory, Docker image, and port configurations:

```
template.hasResourceProperties('AWS::ECS::TaskDefinition', {
      ContainerDefinitions: [
         {
           Environment: [
             {
               Name: Match.exact('NODE_ENV'),
               Value: Match.exact('test'),
             },
           ],
           Essential: true,
           Image: Match.objectEquals({
             'Fn::Sub': Match.stringLikeRegexp('ecr'), // a string that
has ECR on it.
           }),
           LogConfiguration: {
             LogDriver: Match.stringLikeRegexp('awslogs'),
             ...
           },
           Memory: 256,
           Name: Match.exact('Express-test'),
           PortMappings: [
             {
               ContainerPort: 80,
               HostPort: 0,
```

```
                    Protocol: 'tcp',
                  },
              ],
            },
        ],
        ...
    });
```

Let's look at the CloudFormation output for this piece of code, which can be found in the cdk.out/tree.json file. We will see that these values match the CDK CloudFormation output:

```
...
"memory": 256,
"name": "Express-Development",
"portMappings": [
{
    "containerPort": 80,
    "hostPort": 0,
        "protocol": "tcp"
    }
]
...
```

As you can see, assertion tests are, in essence, checklists on "expectations versus reality," with the added detail that we can compare the output of certain constructs or their sub-components. Next, let's look at snapshot tests and see how they compare to assertion testing.

Snapshot tests

Snapshot testing can be useful in a variety of situations, such as when you want to test the output of a complex system or when you want to ensure that the output of a system remains stable over time. It can be especially useful in CDK applications, where it can help you verify that your cloud infrastructure is being deployed correctly and that the resulting resources are in the expected state.

To implement snapshot testing in a CDK application, you need to choose a snapshot testing library that is compatible with the programming language you are using. Some popular snapshot testing libraries for JavaScript and TypeScript include Jest, Ava, and Mocha. In our case, we are using Jest. So, in essence, CDK snapshot tests use the same mechanisms as Jest snapshots, which are predominantly used for the React framework. With Jest snapshots on React, we compare the HTML output of a component with its previous versions; with CDK snapshots, we compare the CloudFormation template, which is the output of our stack, also with its previous versions. Here, the compared output is in the CloudFormation stack definition file, which can be found in the cdk.out directory.

> **Jest's snapshot tests**
>
> Snapshot testing is a feature of the Jest testing framework that's used for UI testing, where the rendered result of a React component is compared to previous versions of it. CDK piggybacks on this feature of Jest, but in this case, the rendered result is CDK's CloudFormation template output. You can find out more by visiting `https://jestjs.io/docs/snapshot-testing`.

As part of the tests we ran, we also executed a snapshot test that's defined in the same `chapter-6-stack.test.ts` file:

```
it('Matches the snapshot.', () => {
    const stack = new Stack();

    const chapter6Stack = new Chapter6Stack(stack, 'Chapter6Stack', {
      env: {
        region: parsed?.CDK_DEFAULT_REGION,
        account: parsed?.CDK_DEFAULT_ACCOUNT,
      },
    });

    const template = Template.fromStack(chapter6Stack);

    expect(template.toJSON()).toMatchSnapshot();
});
```

This is the most basic version of snapshot tests. In the previous block of code, we saved the CloudFormation CDK output in a `template` constant. We then checked that the new output matches the previous snapshot of the CDK application. "Which previous snapshot?" you ask. Well, if you're running these tests for the first time, then no previous snapshot will exist. Due to this, Jest creates one for us. Open the `test/__snapshots__/chapter-6-stack.test.js.snap` file. Let's examine its contents:

```
// Jest Snapshot v1, https://goo.gl/fbAQLP

exports[`Testing Chapter 6 code. Matches the snapshot. 1`] = `
{
  "Mappings": {
    "AWSCloudFrontPartitionHostedZoneIdMap": {
      "aws": {
        "zoneId": "Z2FDTNDATAQYW2",
      },
      "aws-cn": {
        "zoneId": "Z3RFFRIM2A3IF5",
      },
```

```
    },
  },
  "Outputs": {
    "BackendURL": {
      "Value": {
        "Fn::GetAtt": [
          "LBtestD8BBA2D8",
          "DNSName",
        ],
      },
    },
  },
...
```

This output is very familiar. Yes – it's essentially the JSON output of the output CloudFormation stack. Now, let's go ahead and make some changes to the code. Open `lib/chapter-6-stack.ts` and change the ECS task role permissions. Find the following line:

```
actions: ['secretsmanager:GetSecretValue'],
```

Change it to the following:

```
actions: ['secretsmanager:GetSecretValue',
'secretsmanager:ListSecrets'],
```

Now, let's run the tests a gain:

```
$ yarn test
```

Kaboom! We can see that the snapshot test failed, with Jest highlighting exactly where things went wrong. In this case, we added a `secretsmanager:ListSecrets` permission, which was not what Jest expected from looking at the previous snapshot it took:

```
mavdi@Ms-MacBook-Pro-3 infrastructure % yarn test
yarn run v1.22.19
$ yarn build && jest
$ tsc
  FAIL  test/chapter-6-stack.test.js
  Testing Chapter 6 code.
    ✓ The stack has a ECS cluster configured in the right way. (163 ms)
    ✓ The stack has a RDS instance configured in the right way. (92 ms)
    ✓ The stack has the VPC configured in the right way. (86 ms)
    ✗ Matches the snapshot. (94 ms)

  ● Testing Chapter 6 code. › Matches the snapshot.

    expect(received).toMatchSnapshot()

    Snapshot name: `Testing Chapter 6 code. Matches the snapshot. 1`

    - Snapshot  - 1
    + Received  + 4

    @@ -2591,11 +2591,14 @@
          "TaskDefinitiontestTaskRoleDefaultPolicy153B3B8F": {
            "Properties": {
              "PolicyDocument": {
                "Statement": [
                  {
    -               "Action": "secretsmanager:GetSecretValue",
    +               "Action": [
    +                 "secretsmanager:GetSecretValue",
    +                 "secretsmanager:ListSecrets",
    +               ],
                    "Effect": "Allow",
                    "Resource": {
                      "Ref": "MySQLCredentialstestA71485A8",
                    },
                  },

    234 |       const template = Template.fromStack(chapter6Stack);
    235 |
  > 236 |       expect(template.toJSON()).toMatchSnapshot();
        |                                 ^
    237 |   });
    238 | });
    239 |

      at Object.toMatchSnapshot (test/chapter-6-stack.test.ts:236:31)

  › 1 snapshot failed.
Snapshot Summary
  › 1 snapshot failed from 1 test suite. Inspect your code changes or run `yarn test -u` to update them.

Test Suites: 1 failed, 1 total
Tests:       1 failed, 3 passed, 4 total
Snapshots:   1 failed, 1 total
Time:        1.207 s, estimated 2 s
Ran all test suites.
error Command failed with exit code 1.
info Visit https://yarnpkg.com/en/docs/cli/run for documentation about this command.
mavdi@Ms-MacBook-Pro-3 infrastructure % ▉
```

Figure 6.2 – Failing snapshot test

But what if this change was intentional? We might want to list permissions from Secrets Manager in our ECS instance. Well, in that case, once our snapshot test fails (which is what just happened), we review the changes – much like how the `git diff` command works – and if we are OK with the changes, we simply update the snapshot to allow for these changes:

```
$ yarn test --updateSnapshot
```

We've now instructed Jest to take a new snapshot of the CDK output, allowing the changes we just made. Any subsequent runs of `yarn test` will no longer complain about the changes you made.

Snapshot tests with CI/CD

If you're running a CI/CD pipeline (using AWS CodePipeline perhaps), you can add a step where snapshot tests are run as part of the build process. In that case, you need to check for any changes to the snapshot in the Git repository. The pipeline will reject any changes that the developers haven't specifically allowed by also updating the snapshot.

In terms of how the two types of tests compare, in summary, assertion testing compares the actual properties of the created resources with the expected properties, while snapshot testing captures the current state of the stack as a snapshot and compares it with the saved snapshot. Snapshot testing is great at change sensitivity. It helps you review the changes in the stack's output before committing them, whereas assertion tests are more targeted toward construct outputs.

Great – it's safe to think we have a good grasp of various ways to test CDK applications. While testing can get elaborate, you should now have the necessary knowledge to start writing tests for your CDK application.

Examining CDK deployment logs

Another way to troubleshoot a CDK application is by looking at its internal logs. This is often useful if a certain resource gets stuck when being created, which, for example, happens when an ECS container fails to stabilize. In such situations, looking at the verbose CDK logs could shine a light on what is happening.

This can be achieved by passing the following flags to `cdk` commands:

- `-v`: Verbose
- `-vv`: Very verbose
- `-vvv`: Extremely verbose

For instance, when deploying using `cdk deploy`, you can see a lot more information about the status of the deployment if you pass in one of the aforementioned flags, like so:

```
$ cdk deploy --profile cdk -vv
```

There is also the -debug flag. The -debug flag is used to enable debug mode for the cdk command. It will cause the CDK to display additional debug information about the deployment process, including detailed information about the CloudFormation stack events and the individual AWS SDK calls that are being made. This can be useful when you are trying to understand the exact interactions between the CDK and the AWS resources that are being deployed. This flag will also enable the -vvv flag, which will provide even more information about the deployment process.

Go ahead and try some of these flags and see the difference in the amount of output CDK produces for each.

Debugging CDK apps using VSCode

Debugging AWS CDK applications in VSCode allows you to easily identify and fix issues in your code. By using VSCode's built-in debugging tools, such as breakpoints and the ability to step through code, you can quickly locate the source of any errors or bugs and make the necessary changes to resolve them.

Additionally, VSCode's integration with AWS CDK makes it easy to set up and run debugging sessions, allowing you to quickly test and troubleshoot your code without having to switch between multiple tools or environments. Debugging in VSCode can save you time and improve the overall development process of your AWS CDK applications.

> **Debugging**
>
> Code debugging is not a feature that's exclusive to VSCode and there are plenty of resources online on how to set up debugging for other code editors and tools.

To debug AWS CDK using VSCode, you need to set up a launch configuration in your launch. json file. To do that, within the infrastructure folder, create a new directory called .vscode with a launch.json file inside it:

Figure 6.3 – VSCode launch configurations

Inside the launch.json file, make sure you insert the following code:

```
{
    "version": "0.2.0",
    "configurations": [
        {
            "type": "node",
```

```
                    "request": "launch",
        "name": "Launch Program",
        "skipFiles": [
            "<node_internals>/**"
        ],
        "runtimeArgs": [
            "-r", "./node_modules/ts-node/register/transpile-only"
        ],
        "args": [
            "${workspaceFolder}/bin/chapter-6.ts"
        ],
        "env": {
            "CDK_MODE": "ONLY_DEV",
            "CDK_DEFAULT_ACCOUNT": "123456789123",
            "NODE_ENV": "Development",
            "CDK_DEFAULT_REGION": "us-east-1"
        }
            }
        ]
    }
```

Now that we have the debugging configuration in place, we can add breakpoints to our CDK code and access all the helpful tools that come with it.

Let's add a breakpoint to the development stack in the `infrastructure/bin/chapter-6.ts` file:

```
 12
 13   if (['ONLY_DEV'].includes(process.env.CDK_MODE || '')) {
 14     new Chapter6Stack(app, `Chapter6Stack-${process.env.NODE_ENV || ''}`, {
 15       env: { region: 'us-east-1', account: process.env.CDK_DEFAULT_ACCOUNT },
 16     });
 17   }
```

Figure 6.4 – Adding a breakpoint to the CDK code

The breakpoint is that little red dot on the left-hand side of the highlighted line. When the code execution reaches that point, it will pause, allowing you to debug the code just as you would with any other program.

To run the debugger, open its panel by pressing *Ctrl + Shift + D*; you will see a play button labeled **Debug CDK**. You can alter its name by changing the value of the `name` property in the `launch.json` file:

Figure 6.5 – VSCode debugger panel

To start the debugging process, you can either click the play button or press *F5* on your keyboard.

Once the debugging process begins, you can view all the variables in use, both local and global, on the debugging panel:

```
RUN AND DEBUG        ▷  Launch Program   ∨   ⚙  ⋯

∨ VARIABLES                                        ⊟
  ∨ Local
    > __createBinding: ƒ (o, m, k, k2) {
      __dirname: '/home/gabriel/Documents/Programming/W…
      __filename: '/home/gabriel/Documents/Programming/…
    > __importStar: ƒ (mod) {
    > __setModuleDefault: ƒ (o, v) {
    > app: <root>
    > cdk: {alexa_ask: {…}, assertions: {…}, assets: {……
    > chapter_6_pipeline_stack_1: {Chapter6PipelineStac…
    > chapter_6_stack_1: {Chapter6Stack: ƒ, __esModule:…
    > dotenv_1: {config: ƒ, parse: ƒ}
    > exports: {__esModule: true}
    > module: Module {id: '.', path: '/home/gabriel/Doc…
    > require: ƒ require(path) {
    > this: Object
  > Global
```

Figure 6.6 – Variables in the VSCode debugger panel

Additionally, a controller will appear at the top of the screen that provides all the necessary buttons to continue, restart, or end the debugging session:

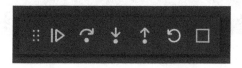

Figure 6.7 – VSCode debugger controls

Let's set two more breakpoints. This time, in the `infrastructure/lib/chapter-6-stack.ts` file, we'll add one for the VPC creation line:

```
36    this.vpc = new Vpc(this, `MyVPC-${process.env.NODE_ENV || ''}`, {
37      cidr,
38      subnetConfiguration: [
39        {
40          cidrMask: 24,
41          name: 'ingress',
42          subnetType: SubnetType.PUBLIC,
43        },
44        {
45          cidrMask: 24,
46          name: 'compute',
47          subnetType: SubnetType.PRIVATE_WITH_NAT,
48        },
49        {
50          cidrMask: 28,
51          name: 'rds',
52          subnetType: SubnetType.PRIVATE_ISOLATED,
53        },
54      ],
55    });
```

Figure 6.8 – Adding a breakpoint for VPC construct initialization

Then, we need to add another one, a little further down, to the RDS construct code:

```
61
62    this.rds = new RDS(this, `RDS-${process.env.NODE_ENV || ''}`, {
63      vpc: this.vpc,
64    });
65
```

Figure 6.9 – Adding the second breakpoint for RDS initialization

We will also keep an eye on two variables: this.vpc and this.rds. This can be done by accessing the **WATCH** tab, in the debugging panel, clicking the + button, and entering the variable names:

```
∨ WATCH                                    +  ⊡  ⊟
    this.vpc: not available
    this.rds: not available
```

Figure 6.10 – Adding the variables to the WATCH tab

Now, if we initiate the debugging process again, once it reaches the this.vpc breakpoint, the variables in the **WATCH** tab will be updated and display all the available data that can be examined during the debugging process:

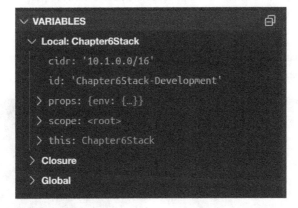

Figure 6.11 – Updated values for the watched variables

However, you may notice that not all values can be viewed. Some of them are represented as ${Token[TOKEN.*]}, which indicates that the value is only determined during deployment time.

The **VARIABLES** tab will also be updated with the local, closure, and global variables available at the current breakpoint:

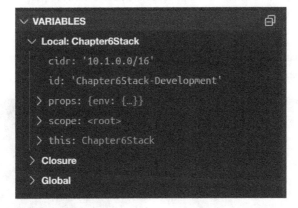

Figure 6.12 – The debugger's VARIABLES tab

By looking at the **variable** tab, you can examine the inner workings of CDK, as well as gain a deeper understanding of the environment surrounding your application, the values you receive, and the values that CDK expects from your code.

Troubleshooting common problems with CDK

As you embark on your journey of coding CDK applications, you may encounter certain challenges that are common among CDK beginners. In this section, we will address some of these challenges and provide effective remedies to overcome them.

Bootstrapping

A common issue when using CDK is deploying to a new environment, which may result in an error stating that the environment has not been bootstrapped. Bootstrapping refers to deploying a pre-created AWS stack to a specific Region. This is necessary to set up the required resources for deploying any stack, S3 buckets, ECR repositories, and so on. As you may recall, we had to deal with this at the beginning of this book.

To resolve this issue, you can use the following command:

```
$ cdk bootstrap
```

You will have to do this everytime to you deploy a CDK project in new AWS accounts.

IAM permissions

When deploying a stack or setting up an environment, AWS resources are created, and it is important to ensure that the account that's used has the appropriate permissions to perform these actions. This includes tasks such as creating an object in S3, uploading a Docker image to ECR, and creating an EC2 instance, among others.

To confirm that your account has the required permissions, go to the AWS console and search for IAM in the top search bar:

Figure 6.13 – Searching for IAM in the AWS console

Once you're in the IAM **Dashboard** area, click on **Users** on the left-hand side panel, find your user, and click on them:

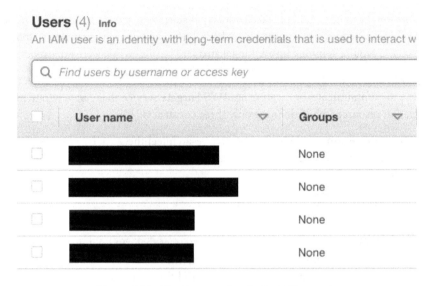

Figure 6.14 – The Users section in the IAM console

This will open a page displaying all the permission policies associated with the account. From here, you can add or remove any necessary permissions to deploy the CDK stack:

Figure 6.15 – Adding the necessary permissions to the IAM user

The easiest to get start policy you can add is the **AdministratorAccess** but you might want to limit the scope of the permissions for the IAM role deploying CDK apps. The AWS IAM best practices are out of the scope of this book but AWS has great resources (for example – `https://aws.amazon.com/blogs/devops/secure-cdk-deployments-with-iam-permission-boundaries/`) on the proper IAM permissions strategy.

Too many resources on the CloudFormation template

As a CDK developer, you may encounter the issue of declaring too many resources in a single stack.

AWS CloudFormation has a hard limit of 500 resources per stack. Going over this limit will result in an error when creating the stack. This limit may seem high, but it can be reached quickly as spinning up one resource can also require spinning up additional auxiliary resources that were not explicitly coded.

For instance, granting access to one resource may result in the creation of IAM objects for the relevant services to communicate, or creating an **AWS Fargate** service that can generate more than 50 CloudFormation resources while defining only three constructs.

To resolve this issue, the recommended approach is to re-architect the stack by reducing the number of resources it contains. This can be achieved by breaking one stack into multiple stacks, using nested stacks, or combining multiple resources into one.

Non-deleted resources

When you delete a CloudFormation stack, by default, resources that can hold user data are kept and not deleted. This means that if you try to redeploy the stack without manually deleting these resources, the deployment will fail due to naming conflicts. To avoid this, you can set the removal policy of resources such as S3 buckets and DynamoDB tables to `destroy` so that they are automatically deleted when the stack is deleted:

```
new s3.Bucket(this, 'Bucket', {
  removalPolicy: cdk.RemovalPolicy.DESTROY,
  });
```

> **Drift errors and failed updates**
>
> On the topic of deleted resources, always make sure you update your stack through CDK. Another common problem is deleting, or updating, a service manually through the AWS console and then later performing an update through CDK. In most cases, the update will fail due to inconsistency in the stack's state, usually referred to as **drift**. More information about this topic can be found here: `https://docs.aws.amazon.com/AWSCloudFormation/latest/UserGuide/using-cfn-stack-drift.html#what-is-drift`.

Summary

In this chapter, we learned about various types of software testing and how testing in CDK is performed using fine-grained assertions and snapshot tests. We learned how, by using them, we can protect sensitive components of CDK applications. We also learned how to examine CDK deployment logs when facing issues. Finally, we learned how to debug our CDK applications in VSCode, as well as troubleshoot common issues. In the next chapter, we will step into the world of serverless development using CDK.

Part 3: Serverless Development with AWS CDK

This part covers exciting ways AWS **Cloud Development Kit** (**CDK**) can be used to make resilient, scalable, and efficient serverless applications without the need for third-party frameworks to deal with the complexity of such stacks. We will show how serverless is where the true power of AWS CDK shines. This part includes the following chapters:

- *Chapter 7, Serverless Application Development with AWS CDK*
- *Chapter 8, Streamlined Serverless Development*

Serverless Application Development with AWS CDK

By this point, you have acquired the skills needed to navigate AWS CDK with ease and can confidently apply CDK to your projects. Throughout this book, we have emphasized that CDK simplifies the maintenance of both traditional cloud applications, whether virtualized or containerized, as well as serverless application development. The challenges of serverless development have previously included the overwhelming amount of JSON and YAML code required to provision various components of a serverless stack. Fortunately, AWS CDK's comprehensive toolkit makes managing these components significantly more manageable. The remaining chapters of this book will delve into various techniques for effectively working with AWS CDK and serverless architecture.

In the previous chapter, we learned how to test our CDK code. In this chapter, we are going to learn how to make our code serverless by ditching **RDS** and **ECS** for **API Gateway** and **AWS Lambda**.

Here's what we'll be covering:

- Explaining what serverless is
- Creating and setting up an API gateway using CDK
- Creating and setting up some Lambda functions and how to connect them to the API gateway
- Creating a **step function** state machine linked to AWS **SES**
- Creating and setting up a **DynamoDB** table and how to automatically populate it during stack deployment

Technical requirements

The source code for this chapter can be found at `https://github.com/PacktPublishing/AWS-CDK-in-Practice/tree/main/chapter-7-serverless-application-development-with-aws-cdk`.

The Code in Action video for this chapter can be viewed at: `https://packt.link/WsOuj`.

Since this chapter is a continuation of the previous one, the directory structure is the same but a few modifications have been made to the services, which we will explore in this chapter.

Configuring the project

Open this chapter's code in your editor of choice. Just like in the previous chapters, we have divided the code into `infrastructure`, `server`, and `web` directories.

First, let's go ahead and build the frontend for our application so that, just like in the previous chapter, CDK can upload the `build` directory's contents to S3. Go to the `web` directory and, in the terminal, run the following command:

```
$ yarn
```

Follow this up with the following command:

```
$ yarn build:prod
```

Then, back in the `infrastructure` directory, run the following command:

```
$ yarn
```

This will install all the project's dependencies.

Also, don't forget to create the `.env.production` file with the required environmental variables.

With that sorted out, let's move on to serverless computing.

What is serverless?

Serverless computing is a way of developing and running cloud-based apps and services without worrying about managing servers. Rather than setting up and maintaining servers as in traditional computing, the cloud provider (such as AWS) handles all the tedious server stuff for you. This means you can focus on writing code and developing your app.

Your app runs in response to events such as a user request or a message in a queue, and the cloud provider automatically handles the code execution and scales things up or down as needed. This saves you time and money while also lowering the risk of downtime and other problems.

As mentioned previously, with serverless computing, you can concentrate on what matters: the functionality of your app. The cloud provider handles all the technical aspects, allowing you to develop faster and take advantage of cutting-edge technology without worrying about compatibility issues.

It's a win-win situation – it is cost-effective, flexible, and scalable, which makes it a popular choice for a lot of companies. It will all make more sense once you jump into it and get coding. So, let's do precisely that.

Creating an API with AWS Lambda and API Gateway

The first thing we need to do is replace the `Express.js` API in ECS and make it serverless. To do that, we are going to use an AWS service called API Gateway.

> **API Gateway**
>
> Amazon API Gateway is a managed service that simplifies the process of creating, publishing, maintaining, monitoring, and securing APIs. It enables developers to create RESTful and WebSocket APIs that can connect to various backend services, including AWS Lambda functions, Amazon Kinesis streams, or any HTTP data source.

In this section, we'll be following these steps to help us create a functioning serverless-powered API for our TODO app:

1. Create and configure the REST API.

2. Create a health check path that's integrated with a Lambda that is going to return `Status 200` and a message stating `OK`.

3. Create the equivalent `GET` and `POST` routes integrated with their respective Lambda function.

If you look in the `infrastructure/lib/constructs/API-GW` directory, you will see an `index.ts` file with the following content:

```
const backEndSubDomain =
  process.env.NODE_ENV === 'Production'
    ? config.backend_subdomain
    : config.backend_dev_subdomain;

const restApi = new RestApi(this, 'chapter-7-rest-api', {
  restApiName: `chapter-7-rest-api-${process.env.NODE_ENV || ''}`,
  description: 'serverless api using lambda functions',
  domainName: {
    certificate: acm.certificate,
    domainName: `${backEndSubDomain}.${config.domain_name}`,
    endpointType: EndpointType.REGIONAL,
    securityPolicy: SecurityPolicy.TLS_1_2,
  },
  deployOptions: {
    stageName: process.env.NODE_ENV === 'Production' ? 'prod' :
'dev',
  },
});
```

This segment of the code generates a RESTful API using a construct from API Gateway. Alongside that, we are configuring the API to utilize the certificate and domain name we created earlier and establishing a stage name in alignment with the deployed environment.

Inside this file, you can also find the following code on line 97:

```
new ARecord(this, 'BackendAliasRecord', {
    zone: route53.hosted_zone,
    target: RecordTarget.fromAlias(new targets.ApiGateway(restApi)),
    recordName: `${backEndSubDomain}.${config.domain_name}`,
});
```

This allows us to use a customized backend subdomain as a DNS alias for the API Gateway URL, as we did with ECS and its load balancer. With our DNS configured, let's create a basic API endpoint.

Creating a health check path

Like the `Express.js` API, we will need a health check path to ensure that the API is operational after deployment by creating a straightforward request.

In the same `index.ts` file, we will be creating a health check path in the recently created REST API. The `addResource()` function creates the path in the API. By default, it only has the root path, which we'll use later:

```
const healthcheck = restApi.root.addResource('healthcheck');
const rootResource = restApi.root;

healthcheck.addMethod('GET', healthCheckLambdaIntegration);
healthcheck.addCorsPreflight({
allowOrigins: ['*'],
allowHeaders: ['*'],
allowMethods: ['*'],
statusCode: 204,
});
```

In this file, we're also creating the method that we want for this path. In this case, it's a GET method connected to a Lambda function. Also, we're setting up the CORS configuration for the same path. The API endpoint has been created, but at this moment, it's not connected to anything that serves the request. Let's go ahead and connect a Lambda function to it so that a response can be generated.

Lambda function integration

API Gateway enables us to use a Lambda function as the code that runs when a request is made to an endpoint. This means the request is passed to the Lambda function and it's up to the function to handle it. We need to return the appropriate status, body, and headers for everything to work correctly. Otherwise, it will return a 500 error code and a message stating Internal Server Error.

Lambda functions

Lambda functions are event-driven, serverless functions that automatically run your code in response to an event, such as an HTTP request, without the need for you to manage the infrastructure. These functions can be used to perform a variety of tasks, such as processing data, handling user requests, or integrating with other services. They are written in popular programming languages, simple to set up, cost-effective, and highly scalable, making them a popular choice for modern applications.

If you look in the `infrastructure/lib/constructs/healthcheck` directory, you'll see another `index.ts` file and a folder called `code`. Inside this folder, we'll store the code that runs inside the Lambda function. Inside `healthcheck/index.ts`, we have the following code:

```
export class HealthCheckLambda extends Construct {
  public readonly func: NodejsFunction;

  constructor(scope: Construct, id: string, props: any) {
    super(scope, id);

    this.func = new NodejsFunction(scope, 'health-check-lambda', {
      runtime: Runtime.NODEJS_16_X,
      entry: path.resolve(__dirname, 'code', 'index.ts'),
      handler: 'handler',
      timeout: Duration.seconds(60),
      environment: {},
      logRetention: logs.RetentionDays.TWO_WEEKS,
    });
  }
}
```

This class is responsible for creating the Lambda function itself. The `NodeJsFunction()` method creates a Lambda function that runs any Node.js code. As you can see, we can configure the Lambda function to our liking. Let's take a look at our configuration:

- `runtime`: The Node.js version you want the Lambda to run.

- `entry`: The path to the code this Lambda function is going to run.

- `handler`: The exported function inside the code specified in the entry property. Every time you run the Lambda function, it will look for this exported function and run everything inside its scope.

- `timeout`: The maximum duration for which the Lambda will run.

- `environment`: The environmental variables you want to pass to the Lambda function. In this case, we are not passing any, but this will be useful later.

- **logRetention**: How long **CloudWatch** is going to retain the Lambda's log. Every time the Lambda runs, it will create a log on its log group in CloudWatch. It is recommended that you set up a retention policy to avoid accumulating these logs, which would incur unwanted and unnecessary costs.

Now, let's examine the code our Lambda is set to execute:

```
export const handler = async () => {
  try {
    return httpResponse(200, JSON.stringify('OK'));
  } catch (error: any) {
    console.error(error);

    return httpResponse(400, JSON.stringify({ message: error.message
}));
  }
};
```

As you can see, is a straightforward code. Its main goal is to check if the API is up and running.

Did you notice the function named `handler`? This is the method the Lambda is going to search for and run, as mentioned previously.

You can also see an `httpResponse()` method. This was created to simplify what was also mentioned previously – to handle every request's return.

Let's take a look at this function:

```
export const httpResponse = (
  statusCode: number,
  body: string,
): IHttpResponse => ({
  body,
  headers: {
    'Access-Control-Allow-Origin': '*',
    'Content-Type': 'application/json',
    'Access-Control-Allow-Methods': 'GET,OPTIONS,POST',

  },
  statusCode,
});
```

It will receive the status code and the body and return everything, the header included.

Now that we have everything we need, let's integrate this Lambda function with the /healthcheck path.

If you go back to the `infrastructure/lib/constructs/API-GW/index.ts` file, you will see the following code:

```
const healthCheckLambda = new HealthCheckLambda(
    this,
    'health-check-lambda-api-endpoint',
    {},
  );

...

const healthCheckLambdaIntegration = new LambdaIntegration(
    healthCheckLambda.func,
  );

...

healthcheck.addMethod('GET', healthCheckLambdaIntegration);
```

Here, we are creating an instance of the Lambda function within the API Gateway's `index.ts` file and using the `LambdaIntegration()` method to enable our Lambda to be integrated with API Gateway.

As mentioned previously, the `healthCheckLambdaIntegration` constant is being passed as a second parameter for the `addMethod()` function.

Congratulations – you just integrated a Lambda function into an API Gateway endpoint!

Completing and deploying the serverless backend

So far, we have successfully created a basic health check endpoint, which has helped you understand how to create Lambda-backed API endpoints. However, the current functionality of our health check endpoint is rather limited, as its name implies. In this section, we will delve deeper into querying and writing data to DynamoDB from the Lambda function handlers, allowing for more complex and useful API endpoints to be built.

Creating a GET and POST route to perform DynamoDB operations

Now that we have our API Gateway properly configured, to perform the same operations and requests we were doing with ECS and RDS, we need to create two more endpoints: one to fetch all data from the DynamoDB table, and another to insert data into it.

In this section, we'll use our previous knowledge of creating an API Gateway method and integrating it with a Lambda function to create two more Lambdas connected to the DynamoDB table:

1. First, let's create the Lambda function we will use to fetch all data from `Table`.

2. Go to `infrastructure/lib/constructs/Lambda/get/code/index.ts`; you will find the following code:

```
export const handler = async () => {
  try {
    const tableName = process.env.TABLE_NAME as string;
    const dynamoDB = new DynamoDB.DocumentClient({
      region: process.env.REGION as string,
    });

    const { Items }: DynamoDB.ScanOutput = await dynamoDB
      .scan({ TableName: tableName })
      .promise();

    return httpResponse(200, JSON.stringify({ todos: Items }));
  } catch (error: any) {
    console.error(error);

    return httpResponse(400, JSON.stringify({ message: error.
message }));
  }
};
```

Let's go through it. Similar to the previous Lambda we created, we are exporting a `handler()` function. Inside it, we have a `tableName` constant that stores the `TABLE_NAME` environment variable and another constant that stores the initialized DynamoDB `DocumentClient()` API. This API will be necessary for us to perform the desired table operation. We are also specifying in which region we want the API to look for a `Table` using another environment variable.

The operation itself is the `scan()` method. As its name implies, this will scan `Table` specified on the `TableName` property and return its items in an array of objects. Once it's returned, we pass this data to the `httpResponse()` handler.

3. Now, we need to create the Lambda that will run this code on AWS.

Go to `infrastructure/lib/constructs/Lambda/get/index.ts`; you will find the following code:

```
const { dynamoTable } = props;

    this.func = new NodejsFunction(scope, 'dynamo-get', {
      runtime: Runtime.NODEJS_16_X,
      entry: path.resolve(__dirname, 'code', 'index.ts'),
```

```
        handler: 'handler',
        timeout: Duration.seconds(60),
        environment: {
          NODE_ENV: process.env.NODE_ENV as string,
          TABLE_NAME: dynamoTable.tableName,
          REGION: process.env.CDK_DEFAULT_REGION as string,
        },
        logRetention: logs.RetentionDays.TWO_WEEKS,
      });

      dynamoTable.grantReadData(this.func);
```

As you can see, it's similar to the code we used to create the health check Lambda, but now, we are using class properties, passing environmental variables to the Lambda, and granting the necessary AWS permission for it to perform the desired operation.

4. With that done, we now need to instantiate this Lambda and integrate it with API Gateway.

    Go to `infrastructure/lib/constructs/API-GW/index.ts`; you will find the following code:

```
const dynamoGet = new DynamoGet(this, 'dynamo-get-lambda', {
    dynamoTable,
  });

...

const dynamoGetIntegration = new LambdaIntegration(dynamoGet.
func);

...

const rootResource = restApi.root;

...

rootResource.addMethod('GET', dynamoGetIntegration);
rootResource.addCorsPreflight({
    allowOrigins: ['*'],
    allowHeaders: ['*'],
    allowMethods: ['*'],
    statusCode: 204,
});
```

Here, we instantiated the `DynamoGet()` construct, which contains the Lambda function we previously saw, using the `LambdaIntegration()` method to integrate it with API Gateway. We also added a `GET` method to the root path, linking it to our Lambda, and added the necessary `CORS` configuration.

5. With that done, we now need to create the Lambda function for the POST request to insert an item into our `Table`.

Go to `infrastructure/lib/constructs/Lambda/post/code/index.ts`; you will find the following code:

```
export const handler = async (event: PostEvent) => {
  try {
    const { todo_name, todo_description, todo_completed } =
JSON.parse(
      event.body,
    ).todo;
    const tableName = process.env.TABLE_NAME as string;
    const dynamoDB = new DynamoDB.DocumentClient({
      region: process.env.REGION as string,
    });

    const todo: Todo = {
      id: uuidv4(),
      todo_completed,
      todo_description,
      todo_name,
    };

    await dynamoDB.put({ TableName: tableName, Item: todo
}).promise();

    return httpResponse(200, JSON.stringify({ todo }));
  } catch (error: any) {
    console.error(error);

    return httpResponse(400, JSON.stringify({ message: error.
message }));
  }
};
```

Similar to the GET request Lambda code, we instantiate the `DocumentClient()` API to perform the desired operation on the table – in this case, the PUT operation. We are also destructuring the request body to get the necessary data to create our item. We are also creating a constant to store the object that will be inserted into our `Table`, for organization purposes. The rest of the code follows the same logic as the previous Lambda.

---

try/catch

It is important to note that this and all previous Lambda codes are wrapped in a `try/catch` statement, meaning that any errors will be handled by the code inside the `catch` block.

---

6.   The next step is to create the Lambda function. Go to `infrastructure/lib/constructs/Lambda/post/index.ts`; you will find the following code:

```
const { dynamoTable } = props;

  this.func = new NodejsFunction(scope, 'dynamo-post', {
    runtime: Runtime.NODEJS_16_X,
    entry: path.resolve(__dirname, 'code', 'index.ts'),
    handler: 'handler',
    timeout: Duration.seconds(60),
    environment: {
      NODE_ENV: process.env.NODE_ENV as string,
      TABLE_NAME: dynamoTable.tableName,
      REGION: process.env.CDK_DEFAULT_REGION as string,
    },
    logRetention: logs.RetentionDays.TWO_WEEKS,
  });

  dynamoTable.grantWriteData(this.func);
```

This code follows the same logic as the Lambda we created for the GET request. The only differences are the code that runs inside the Lambda and the permission to write data into our Table.

7.   Now, let's take a look in the `infrastructure/lib/constructs/API-GW/index.ts` file, where we will find the following code:

```
const dynamoPost = new DynamoPost(this, 'dynamo-post-lambda', {
    dynamoTable,
  });

...

const dynamoPostIntegration = new LambdaIntegration(dynamoPost.
func);

...

rootResource.addMethod('POST', dynamoPostIntegration);
```

This code also follows the same logic we used for the GET request. We instantiated the Lambda, used the LambdaIntegration() method to integrate it with API Gateway, and added a POST method to the root path that is linked to our Lambda.

Simplifying the Lambda integration

In addition to the method shown earlier, there is an even simpler way to integrate with AWS Lambda using a construct called `LambdaRestApi`. This function sets up the REST API with a default Lambda function as its handler and a path that uses the greedy proxy option of (`"{proxy+}"`) and the `ANY` method from API Gateway. This means that every request will be automatically routed to this Lambda function, and the function will be responsible for processing the request, instead of having to create a resource and method for each request as we did.

Here's an example of how you can set up this constructor:

```
new LambdaRestApi(this, 'MyRestApi', {
  handler: lambda.func,
  restApiName: 'rest-api-name'
  defaultCorsPreflightOptions: {
    allowOrigins: ['*'],
    allowHeaders: ['*'],
    allowMethods: ['*'],
    statusCode: 204,
  },
});
```

Yep, it's that simple. In this example, the main differences are the new `handler` property, the `default` lambda function, the CORS configuration, which we previously had to do individually for each path, and the missing DNS configuration, which is optional.

Although this seems quite practical, it has its downsides. Inside this default `Lambda` function, you will have to deal with every incoming request, just like you need to do in an Express.js API, for example. But this level of convenience can vary from project to project.

Great – it's coming along nicely. You might have noticed that we haven't set up a DynamoDB table yet. Let's take a look at how to do that now.

Creating and configuring a DynamoDB instance

With the Lambdas and our API Gateway created, the only remaining step is to replace RDS with DynamoDB by creating a table.

> **DynamoDB**
> DynamoDB is a fully managed, highly scalable, and low-latency NoSQL database service provided by **Amazon Web Services (AWS)**. It allows you to store, retrieve, and update data in a flexible, JSON-like format. DynamoDB is designed to handle high read and write throughput and can automatically scale up or down based on usage. It supports both document and key-value data models, and it can be easily integrated with other AWS services.

If you navigate to `infrastructure/lib/constructs/DynamoDB/index.ts`, you will find the following code:

```
this.table = new Table(this, `Dynamo-Table-${process.env.NODE_ENV ||
''}`, {
      partitionKey: { name: 'id', type: AttributeType.STRING },
      tableName: `todolist-${process.env.NODE_ENV?.toLowerCase() ||
''}`,
      billingMode: BillingMode.PAY_PER_REQUEST,
      removalPolicy: RemovalPolicy.DESTROY,
    });

    new DynamoDBSeeder(
      this,
      `Dynamo-InlineSeeder-${process.env.NODE_ENV || ''}`,
      {
        table: this.table,
        seeds: Seeds.fromInline([
          {
            id: uuidv4(),
            todo_name: 'First todo',
            todo_description: "That's a todo for demonstration
purposes",
            todo_completed: true,
          },
        ]),
      },
    );
```

This part of the code is responsible for creating the DynamoDB table and automatically seeding it during deployment time. The `Table()` construct configures the table:

- `partitionKey`: This is much like the primary key of a relational database. You need to specify a key that every item in that database needs to have. In this case, since we are replicating the type of data that was on RDS, we need an `id` key of the `string` type.

- `tableName`: This is the name of the table.

- `billingMode`: This shows how the read and write table operations are going to be charged.

- `removalPolicy`: This defines what happens to the table if it stops being managed by **CloudFormation**. The default option is to retain the resource.

Next to it, we have the `DynamoDBSeeder()` construct. Similar to how we were populating RDS with the `script.sql` file, we are going to use this construct for the same purpose.

DynamoDB Seeder is a third-party construct, and you can read more about it on its npm page: `https://www.npmjs.com/package/@cloudcomponents/cdk-dynamodb-seeder/v/2.1.0`.

The way this construct works is quite simple. As you can see in the code, you specify the table and the data you want to insert into the `seeds` property. You can use a seed from a JSON file, a S3 bucket, or, as in this case, an inline array of objects.

Lastly, we need to include the API Gateway and DynamoDB constructs in our stack. This will allow us to deploy these resources to the cloud and make them available for use.

If you take a look at the `infrastructure/lib/chapter-7-stack.ts` file, you will see the code that is responsible for creating the stack and adding the API Gateway and DynamoDB constructs to it:

```
this.dynamo = new DynamoDB(this, `Dynamo-${process.env.NODE_ENV ||
''}`);

new ApiGateway(this, `Api-Gateway-${process.env.NODE_ENV || ''}`, {
    route53: this.route53,
    acm: this.acm,
    dynamoTable: this.dynamo.table,
});
```

Here, we are adding the API Gateway and DynamoDB constructs to our CDK stack. Since we no longer need the ECS, RDS, and VPC constructs, we can remove them from the Stack.

We are also passing the Route53, ACM, and DynamoDB table constructs to API Gateway so that it can create, pass, and configure the necessary data for those resources, as we discussed earlier.

Great job! You've now built a fully serverless API endpoint that connects to a Lambda function and queries a DynamoDB table. With this knowledge, you're now equipped to start developing serverless applications. However, as you progress to more complex applications, you'll find that managing routines and application state becomes increasingly challenging. That's where AWS Step Functions comes in. In the next section, we'll explore how Step Functions can help you manage the complexity of your serverless applications, allowing you to focus on building the core logic of your application.

Introduction to Step Functions

AWS Step Functions is a workflow management service provided by AWS that enables you to create, execute, and visualize multistep operations or applications.

Think of Step Functions as a tool to streamline complex operations in a serverless environment by connecting and coordinating different tasks. These tasks can range from AWS Lambda functions to AWS Batch jobs to AWS Glue jobs. With Step Functions, you no longer have to manually perform each step of a workflow; instead, you just define the flow of your operations in a state machine and let Step Functions do the heavy lifting for you.

One of the key advantages of using step functions is the ability to keep track and provide a visual representation of the workflow. This makes it easier to understand what's happening, since you can see exactly what's going on and where things went wrong, making debugging and troubleshooting easier.

Step functions are widely used in applications such as data pipelines, big data processing, infrastructure management, and application deployment. They're an essential component of many serverless architectures, allowing you to build complex serverless apps without having to manage servers or infrastructure.

With that in mind, let's incorporate a step function into our app. The goal will be to trigger the execution of a state machine every time we hit a certain route. Once the flow is complete, we will receive an email indicating the source of the trigger.

Let's kick things off by creating our `Step Function` construct:

1. Go to `infrastructure/lib/constructs/Step-Function/index.ts`; you will find the code for creating the state machine and configuring the necessary step task to send an email. Let's dive into the code step by step.

    Right off the bat, we have the following piece of code:

    ```
    const emailAddress = process.env.EMAIL_ADDRESS;
    const resourceArn = `arn:aws:ses:${Stack.of(this).region}:${
      Stack.of(this).account
    }:identity/${emailAddress}`;

    const verifyEmailIdentityPolicy = AwsCustomResourcePolicy.
    fromStatements([
      new PolicyStatement({
        actions: ['ses:VerifyEmailIdentity', 'ses:DeleteIdentity'],
        effect: Effect.ALLOW,
        resources: ['*'],
      }),
    ]);
    new AwsCustomResource(this, `Verify-Email-Identity-${process.
    env.NODE_ENV || ''}`, {
      onCreate: {
        service: 'SES',
        action: 'verifyEmailIdentity',
        parameters: {
          EmailAddress: emailAddress,
        },
        physicalResourceId: PhysicalResourceId.of(`verify-
    ${emailAddress}`),
        region: Stack.of(this).region,
      },
      policy: verifyEmailIdentityPolicy,
      logRetention: 7,
    });
    ```

2. In this part of the code, we set up the SES email identity. Keep in mind that you'll need to add an email address to the `.env` file you're using. For this example, the recipient will also be the sender, meaning you'll be sending an email to yourself. This is just for demonstration purposes. In reality, you'd have an email address such as `noreply@test.com` to send these types of emails.

This step is crucial so that SES can send an email using the designated email address. Once you deploy the stack, you'll receive a verification email from Amazon.

A little further down the file, you will find the following code:

```
const emailBody =
'<h2>Chapter 7 Step Function.</h2><p>This step function was
triggered by: <strong>{}</strong>.';

const sendEmail = new CallAwsService(this, `Send-Email-
${process.env.NODE_ENV || ''}`, {
  service: 'sesv2',
  action: 'sendEmail',
  parameters: {
    Destination: {
      ToAddresses: [emailAddress],
    },
    FromEmailAddress: emailAddress,
    Content: {
      Simple: {
        Body: {
          Html: {
            Charset: 'UTF-8',
            Data: JsonPath.format(
          emailBody,
              JsonPath.stringAt('$.message'),
            ),
          },
        },
        Subject: {
          Charset: 'UTF-8',
          Data: 'Chapter 7 Step Function',
        },
      },
    },
  },
  iamResources: [resourceArn],
});
```

3. In this section of the code, we set up the first and only step in our state machine. This step will directly link to SES using its ARN. Take note of how we separate the email body into a constant. The body must be a string in HTML format, and the curly brackets indicate that we expect a dynamic value at that position, specified in the `JsonPath.stringAt('$.message')` function. When the state machine is triggered, we pass an object containing a property named `message` with whatever message we want, in this case indicating where the step function was triggered from.

    With this configuration done, let's move on to our `stateMachine`:

```
const stateMachine = new StateMachine(this, 'State-Machine', {
  definition: sendEmail,
  timeout: Duration.minutes(5),
});

stateMachine.role.attachInlinePolicy( new Policy(this,
`SESPermissions-${process.env.NODE_ENV || ''}`, {
  statements: [
    new PolicyStatement({
      actions: ['ses:SendEmail'],
      resources: [resourceArn],
    }),
  ],
}));
this.stateMachine = stateMachine;
```

    In this part, we are configuring the state machine. It's pretty straightforward – all we need to do is define the flow in the `definition` and set a timeout in case something goes wrong and the machine gets stuck. Don't forget to attach a policy to the state machine, detailing the actions it's allowed to perform. Lastly, we must store the state machine as a variable for future reference in the stack.

4. The next step is to add the necessary code on the Lambda. Go to `infrastructure/lib/constructs/Lambda/get/code/index.ts`; you will see the following code:

```
const stepFunctions = new StepFunctions({
  region: process.env.REGION as string,
});
...
await stepFunctions.startExecution({
  stateMachineArn: process.env.STATE_MACHINE_ARN as string,
  input: JSON.stringify({
    message: 'GET / route',
  }),
}).promise();
```

You'll find the following code in .../Lambda/post/code/index.ts:

```
const stepFunctions = new StepFunctions({
  region: process.env.REGION as string,
});
...
await stepFunctions.startExecution({
  stateMachineArn: process.env.STATE_MACHINE_ARN as string,
  input: JSON.stringify({
    message: 'POST / route',
  }),
}).promise();
```

Here, we are using the Step Function API to start execution by calling the startExecution() method. Then, we pass stateMachineARN, obtained from an environmental variable, and the desired input message to the method. Both codes implement the same logic, with the only difference being the input message that's sent when triggering the state machine.

5. Now, all that's left is to add the construct to the stack and pass the state machine stored in a variable, as mentioned previously, to API Gateway.

    Go to infrastructure/lib/chapter-7-stack.ts; you will see this code:

```
this.stepFunction = new StepFunction(this, `Step-Function-
${process.env.NODE_ENV || ''}`, {});

new ApiGateway(this, `Api-Gateway-${process.env.NODE_ENV ||
''}`, {
  route53: this.route53,
  acm: this.acm,
  dynamoTable: this.dynamo.table,
  stateMachine: this.stepFunction.stateMachine,
});
```

Inside the API Gateway construct, we are passing the state machine to both Lambdas:

```
const { acm, route53, dynamoTable, stateMachine } = props;
...
const dynamoPost = new DynamoPost(this, 'dynamo-post-lambda', {
  dynamoTable,
  stateMachine,
});
const dynamoGet = new DynamoGet(this, 'dynamo-get-lambda', {
  dynamoTable,
  stateMachine,
});
```

Inside the `DynamoPost()` and `DynamoGet()` functions, we are passing our stateMachineARN as an environment variable and granting the necessary permissions to the Lambda:

```
this.func = new NodejsFunction(scope, 'dynamo-post', {
  runtime: Runtime.NODEJS_16_X,
  entry: path.resolve(__dirname, 'code', 'index.ts'),
  handler: 'handler',
  timeout: Duration.seconds(60),
  environment: {
    NODE_ENV: process.env.NODE_ENV as string,
    TABLE_NAME: dynamoTable.tableName,
    REGION: process.env.CDK_DEFAULT_REGION as string,
    STATE_MACHINE_ARN: stateMachine.stateMachineArn,
  },
  logRetention: logs.RetentionDays.TWO_WEEKS,
});
dynamoTable.grantWriteData(this.func);
stateMachine.grantStartExecution(this.func);
...
this.func = new NodejsFunction(scope, 'dynamo-get', {
  runtime: Runtime.NODEJS_16_X,
  entry: path.resolve(__dirname, 'code', 'index.ts'),
  handler: 'handler',
  timeout: Duration.seconds(30),
  environment: {
    NODE_ENV: process.env.NODE_ENV as string,
    TABLE_NAME: dynamoTable.tableName,
    REGION: process.env.CDK_DEFAULT_REGION as string,
    STATE_MACHINE_ARN: stateMachine.stateMachineArn,
  },
  logRetention: logs.RetentionDays.TWO_WEEKS,
});
dynamoTable.grantReadData(this.func);
stateMachine.grantStartExecution(this.func);
```

Once completed, you can deploy the production stack and see your serverless application running on AWS by running the following command:

```
$ yarn cdk deploy --profile cdk
```

Summary

In this chapter, we created and integrated Lambda functions and an API gateway to replace the RDS and ECS services. We created a DynamoDB table and set up a way to automatically seed it when deploying the stack and created a step function state machine linked to SES. We also set up a GET request to fetch all data from the DynamoDB table and a POST request to insert data into the same table and trigger the state machine. Finally, we added the API Gateway and DynamoDB constructs to our CDK stack and removed the ECS, RDS, and VPC constructs from the Stack. As you've experienced, the process of applying changes and testing serverless logic can be time-consuming. In the next chapter, we'll explore ways to streamline building serverless applications, specifically by optimizing our local environment for faster and more efficient development.

Streamlined Serverless Development

Previously, we covered how to convert our code into a serverless architecture by replacing ECS, RDS, and VPC services with API Gateway, Lambda functions, and DynamoDB. But there's still room for improvement in the development process.

Even though using serverless technologies simplifies maintenance after deployment, we can still optimize the way we work on the code locally.

In this chapter, we'll be covering the following topics:

- Common problems with serverless development
- Running Lambda application logic locally and integrating it with a local express server
- How to run AWS services locally using the **LocalStack** toolset

By the end of this chapter, you will be aware of the problems in serverless-oriented development, and your code will have been streamlined to the point where you will be able to develop **infrastructure as code (IaC)** and test it locally before deploying it to AWS.

Technical requirements

The source code for this chapter can be found at `https://github.com/PacktPublishing/AWS-CDK-in-Practice/tree/main/chapter-8-streamlined-serverless-development`.

The Code in Action video for this chapter can be viewed at: `https://packt.link/06ppp`.

Since it continues from the previous chapter, the directory structure is the same, with a few modifications to the services that we will explore in this chapter.

Configuring the project

Open the chapter code in your editor of choice. Just like in the previous chapters, we have divided up the code into the `infrastructure`, `server`, and `web` directories.

First, let's go ahead and install all the dependencies for our `infrastructure` and `server` folders. Go to the `infrastructure` directory and in the terminal, run the following:

```
$ yarn
```

Then, back in the `server` directory, run the following command:

```
$ yarn
```

This will install all the project's dependencies.

With that sorted out, let's move on.

Common problems with serverless development

Developing serverless applications has become a popular and useful tool in the tech industry, promising to simplify the deployment and creation of applications. However, as with any new technology, there are always a few bumps in the road.

One of the most common issues is a lack of visibility into the underlying infrastructure. Unlike traditional server-based applications, where the servers and their configurations are visible and accessible, serverless applications abstract these away, making it more difficult to identify and solve issues when they arise.

Another problem is the lack of control over the environment. As serverless technologies are designed to abstract away the underlying infrastructure, developers are often limited in the amount of control they have over the environment in which their applications run. This makes it challenging to implement certain types of functionality, such as fine-grained access control or data storage, and can also make it difficult to integrate with existing systems and tools. Platforms such as AWS keep adding more options for developers to run their code in environments they can customize. Custom Lambda container images, where AWS allows you to create your own Lambda runtimes over the open source base images (more information can be found at `https://docs.aws.amazon.com/lambda/latest/dg/images-create.html`), provide us with a step in the right direction.

Debugging can also be a challenge in serverless development, particularly if your code is running on multiple AWS Lambda functions. This can make it difficult to track down issues and find their root cause, so it's crucial to have logs properly set up.

Testing is another area where serverless development can be tricky. In a traditional development environment, you can run tests locally on your own machine, but with serverless development, you're usually relying on the cloud to run your tests, which can lead to slow test runs and increased costs.

Also, when coding serverless applications, developers lack a proper local development environment. This can be a real pain because it means you have to deploy your code to a live environment just to test it, which can be both time-consuming and costly.

Despite these challenges, developers continue to embrace serverless development due to the numerous benefits it provides, such as scalability and cost savings as previously discussed.

In *Chapter 10* of the book, we will dive into more open source alternatives to the AWS platform for serverless development, which might help overcome some of these limitations.

Running Lambda application logic locally

You might have noticed that since we changed the infrastructure code to serverless, we haven't touched the server's folder code. You might be wondering if we are still going to use it. The answer is yes: we are going to use it for local development in the same way we were using it when we had RDS, but we will have to make some changes to make it work with DynamoDB. Additionally, we will import the same code that is running inside the deployed Lambda function into the local server and run it in their POST and GET routes. So, let's get started.

You can find all the code discussed here in the `server/src/index.ts` file.

The first thing we need to do is import the code used in the Lambda integration of API Gateway for POST and GET endpoints:

```
import { handler as PostHandler } from '../../infrastructure/lib/
constructs/Lambda/post/lambda';
import { handler as GetHandler } from '../../infrastructure/lib/
constructs/Lambda/get/lambda';
```

Since both functions are exporting `handler()` <GetHandler>, we are using `Import Aliases` to differentiate each function. With that done, let's move on to the `Express` code itself. In the POST route, we are going to delete everything from it and replace it with the `PostHandler()` function:

```
app.post('/', async (req, res) => {
    const event = {
      body: JSON.stringify(req.body),
    };

    const { statusCode, body } = await PostHandler(event);

    return res.status(statusCode).send(body);
  });
```

Also, remember the `httpResponse()` function we're using? Its output will always look like the following:

```
export interface IHttpResponse {
  body: string;
  headers: {
    'Access-Control-Allow-Origin': string;
    'Content-Type'?: string;
    'Access-Control-Allow-Headers'?: string;
    'Access-Control-Allow-Methods'?: string;
    'X-Requested-With'?: string;
  };
  statusCode: number;
}
```

In order for the API to work properly, we need to take its output and pass it to the `Express res()` method. To make it more elegant, we're destructuring `PostHandler()` and returning the status code and body.

Next, we need to change the `GET` route:

```
app.get('/', async (_req, res) => {
    const { statusCode, body } = await GetHandler();

    return res.status(statusCode).send(body);
});
```

As you can see, the logic is similar to that of the `POST` route, but with a few small changes.

We also changed the `/healthcheck` path to match the route name and output with those of API Gateway:

```
app.get('/healthcheck', async (_req, res) => {
    return res.status(200).send(JSON.stringify('OK'));
});
```

If you were to run the code now, with the proper environment variables set, you'd see an error when trying to access the DynamoDB table in the request. The reason is that the code is trying to reach a table on AWS that doesn't exist. Although you could spin up a subset of the stack, with only the DynamoDB construct and the local server running, and see the API working as expected, that's not an ideal solution.

To solve that issue, we'll use a toolset called LocalStack.

Using LocalStack to simulate AWS services

Remember when we said that the lack of a proper local development environment is a bit of a pain when developing serverless applications? To fix that, instead of deploying our code to AWS every time we want to test it, we can use LocalStack to run it locally. LocalStack allows us to mimic the functionality of AWS services, such as DynamoDB and S3, on our own machine. This way, we can test and develop our cloud and serverless apps offline.

Installing LocalStack

To use LocalStack, we first need to install its **command-line interface** (**CLI**). The following is a detailed set of instructions:

1. To be able to install the CLI, you need to have the following installed on your computer:

    *   Python (3.7 up to 3.10)

    *   Pip (Python package manager)

    *   Docker

2. Once you have those set up, you can proceed to your terminal and run the following command:

    ```
    $ python3 -m pip install localstack
    ```

3. If everything goes well, you should see this prompt upon the command's completion:

    ```
    Successfully installed localstack-1.3.1
    ```

4. To confirm that LocalStack has been installed, you can run this command in your terminal:

    ```
    $ localstack —help
    ```

5. If you see the following prompt, you're good to go:

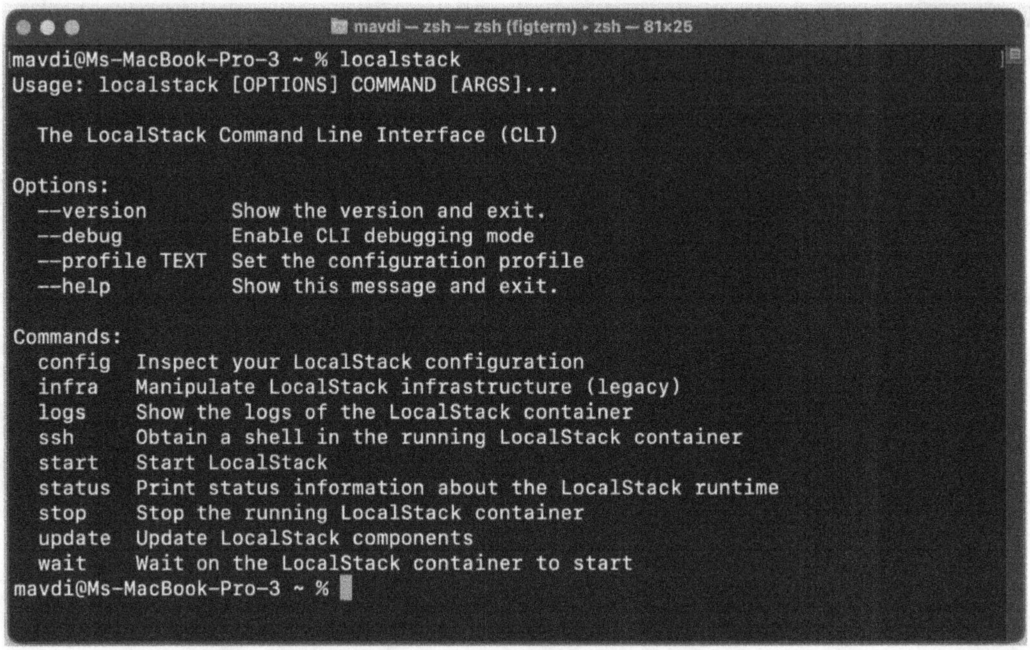

Figure 8.1 – LocalStack command output

> **Note**
>
> In some Linux environments, by default, PATH will not recognize the location where Pip installs its packages. So, you need to manually add that to PATH by adding export PATH=$HOME./ local/bin:$PATH to your .bashrc or .zshrc file.

Starting LocalStack

Now that it's installed, we can start LocalStack by running this command:

```
$ localstack start
```

Once it's finished running, you should see this prompt with Ready. at the end:

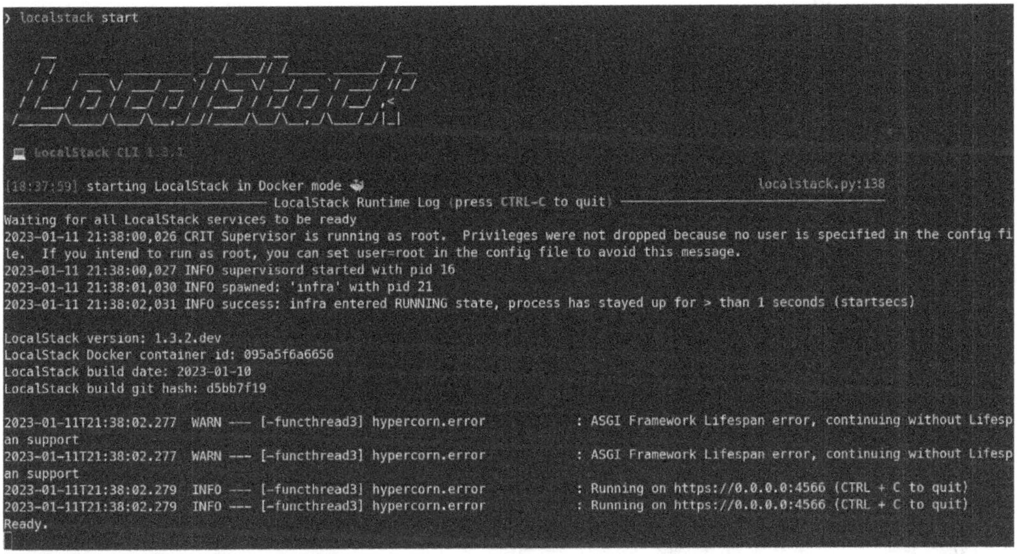

Figure 8.2 – Expected LocalStack command output

You can check whether LocalStack was started correctly by running the `docker container ls` command in another terminal window. You should see the LocalStack container listed:

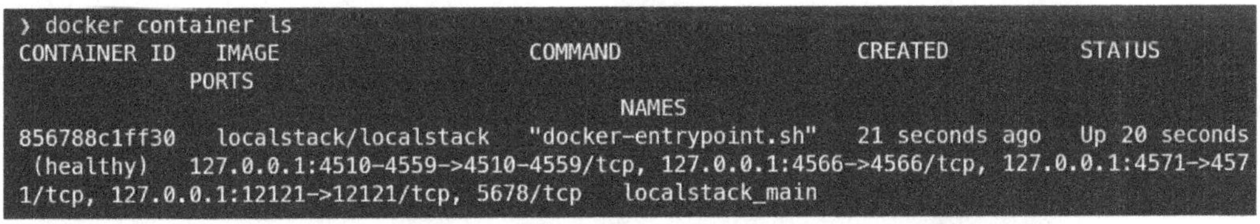

Figure 8.3 – Docker command listing all the existing containers

> **Note**
>
> To learn about alternative methods for installing LocalStack, check out their documentation website: `https://docs.localstack.cloud/getting-started/installation/`.

With that taken care of, our next step is to install `cdklocal`. This is a CLI wrapper required to deploy the CDK code to LocalStack using its APIs. This can be done by installing it globally, as an npm library, by running the following command:

```
$ npm install -g aws-cdk-local
```

To ensure it was installed properly, you can run this command:

```
$ cdklocal --version
```

Afterward, you should see a prompt that looks like the following:

```
) cdklocal --version
2.59.0 (build b24095d)
```

Figure 8.4 – Current cdklocal version

Now, with LocalStack and `cdklocal` installed and running, navigate to the project's `infrastructure` folder. We are going to put some logic in the `lib/chapter-8-stack.ts` file that, if using LocalStack, will deploy only DynamoDB.

Your code will look like the following:

```typescript
export class Chapter8Stack extends Stack {
  public readonly acm: ACM;

  public readonly route53: Route53;

  public readonly s3: S3;

  public readonly vpc: Vpc;

  public readonly dynamo: DynamoDB;

  constructor(scope: Construct, id: string, props?: StackProps) {
    super(scope, id, props);

    const isCDKLocal = process.env.NODE_ENV === 'CDKLocal';

    this.dynamo = new DynamoDB(this, `Dynamo-${process.env.NODE_ENV ||
''}`);

    if (isCDKLocal) return;

    this.route53 = new Route53(this, `Route53-${process.env.NODE_ENV
|| ''}`);

    this.acm = new ACM(this, `ACM-${process.env.NODE_ENV || ''}`, {
      hosted_zone: this.route53.hosted_zone,
    });

    this.s3 = new S3(this, `S3-${process.env.NODE_ENV || ''}`, {
```

```
      acm: this.acm,
      route53: this.route53,
    });


    new ApiGateway(this, `Api-Gateway-${process.env.NODE_ENV || ''}`,
  {
      route53: this.route53,
      acm: this.acm,
      dynamoTable: this.dynamo.table,
    });
  }
}
```

In the terminal, within the `infrastructure` folder, run the following command:

```
$ yarn cdklocal bootstrap
```

Next, run the following command:

```
$ yarn cdklocal deploy
```

The first command will bootstrap the environment on your local cloud, while the second command will deploy the stack into the same local cloud. The output of the second command should look similar to this:

```
✅   Chapter8Stack-Development

✦   Deployment time: 5.41s

Stack ARN:
arn:aws:cloudformation:us-east-1:000000000000:stack/Chapter8Stack-Development/da2092cb

✦   Total time: 10.04s


Done in 11.51s.
```

Figure 8.5 – Message displayed after a stack was successfully deployed

Congratulations! You just deployed the development stack into your local cloud infrastructure.

> **Note**
>
> Keep in mind that LocalStack is ephemeral, meaning if you run the `localstack stop` command, any stack that was deployed previously will be destroyed.

Configuring DynamoDB to work with LocalStack

By default, the LocalStack endpoint is on port 4588 on your localhost. To make sure DynamoDB is reaching the LocalStack endpoint, rather than the default AWS one, we need to configure its endpoint:

1. First, go to `infrastructure/lib/constructs/Lambda/post/lambda/index.ts` where you can find the following code:

```
const tableName = process.env.TABLE_NAME!
const awsRegion = process.env.REGION || 'us-east-1';

const dynamoDB = new DynamoDB.DocumentClient({
  region: awsRegion,
  endpoint:
  Process.env.DYNAMODB_ENDPOINT || `https://
dynamodb.${awsRegion}.amazonaws.com`,
});
```

   Here, we are passing a custom endpoint to the `DocumentClient()` DynamoDB API from a `.env` file that we'll create later. Notice the inline logic in this property; this value will only be present in the `.env` file when running the local server. When deploying the stack to AWS, the default URL will be passed to the `DocumentClient()` DynamoDB API.

2. Now, navigate to `/infrastructure/lib/constructs/Lambda/get/lambda/index.ts` and make the same changes to `DocumentClient()` there:

```
const dynamoDB = new DynamoDB.DocumentClient({
  region: awsRegion,
  endpoint:
  process.env.DYNAMODB_ENDPOINT || `https://
dynamodb.${awsRegion}.amazonaws.com`,
});
```

3. The final step is to create an environment file containing all the necessary information to run our local server.

   Go to the `server/` folder and create a `.env` file at the root of the folder with the following variables and values:

```
PORT=3000
REGION=us-east-1
TABLE_NAME=todolist-cdklocal
DYNAMODB_ENDPOINT=http://localhost:4566
```

   In the terminal, inside the `server` folder, run the following command:

```
$ export AWS_PROFILE=cdk
```

Although it is not mandatory to include your real access key ID and secret access key, LocalStack needs a string value in those properties to simulate the AWS calls. Instead of adding these values to the `.env` file, you can export your AWS profile as we did in previous chapters. Otherwise, LocalStack will produce an error saying `Missing credentials in config`.

4. Still in the `server/` folder, in your terminal, run the following command:

    ```
    $ yarn dev
    ```

5. Once the command is executed, the local development server will start and you should see a message similar to the following, indicating that the server is running:

```
> yarn dev
yarn run v1.22.19
$ tsnd src/index.ts
[INFO] 18:40:05 ts-node-dev ver. 2.0.0 (using ts-node ver. 10.8.2, typescript ver. 4.8.4)
Server is listening on port 3000
```

Figure 8.6 – Message indicating the local server is operational on a specified port

6. You can then test the server by making requests to `localhost:3000`, for example, by sending a request to the `/healthcheck` path to check whether the server is working properly:

```
> curl --request GET \
    --url http://localhost:3000/healthcheck
"OK"%
```

Figure 8.7 – Health check path response

7. Another example is hitting the `POST /` endpoint to create a table item:

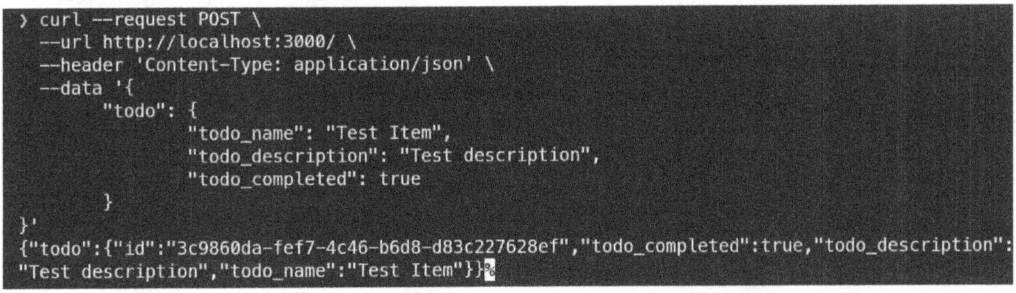

```
> curl --request POST \
    --url http://localhost:3000/ \
    --header 'Content-Type: application/json' \
    --data '{
        "todo": {
                "todo_name": "Test Item",
                "todo_description": "Test description",
                "todo_completed": true
        }
}'
{"todo":{"id":"3c9860da-fef7-4c46-b6d8-d83c227628ef","todo_completed":true,"todo_description":
"Test description","todo_name":"Test Item"}}
```

Figure 8.8 – POST root path response

8.   You can also hit the GET / endpoint to retrieve all items from the table:

```
> curl --request GET \
  --url http://localhost:3000/
{"todos":[{"todo_name":"Test Item","id":"3c9860da-fef7-4c46-b6d8-d83c227628ef","todo_descripti
on":"Test description","todo_completed":true}]}
```

Figure 8.9 – GET root path response

Great – we are all set. Our serverless application is running locally and we have massively reduced the amount of time it takes for us to see the result of our changes since we ran the entire stack locally.

Limitations of LocalStack

We dislike it when a solution to a certain problem is proposed without covering the limitations of that solution. All of that is left for the developers to find out themselves (looking at you, AWS docs) through trial and error, potentially wasting hours or days on a solution that might not be the right one for them.

We've presented LocalStack as a silver bullet for local serverless development, but it too comes with limitations that you might find annoying or impossible to work with. In our company, we use LocalStack on some projects and not on others, mainly depending on the AWS services we are using.

One of the main limitations is that LocalStack may not fully replicate the behavior of some AWS services. For example, the behavior of the S3 service in LocalStack may differ from that of AWS's actual S3 service, which can lead to unexpected errors or behavior when the code is deployed to production. Additionally, not all AWS services are supported by LocalStack, so certain CDK constructs may not be able to be fully tested locally. This is also evident when working with certain versions of AWS Lambda containers for Node.js. Right now, only Node.js v14.x.x is supported by LocalStack.

AWS's official solution for the local development of CDK applications is to use the official **AWS SAM** toolkit. AWS SAM is an open source framework for building serverless applications. It provides a simplified way of defining the Amazon API Gateway APIs, AWS Lambda functions, and Amazon DynamoDB tables needed by your serverless application. By using AWS SAM, you can define your application's resources using YAML or JSON (ugh, so 2008) templates, which are easier to read and maintain than traditional CloudFormation templates.

The SAM CLI toolkit can be used to read the configurations of a Lambda function and have it invoked locally as such:

```
# Invoke MyFunction from the TODOStack
sam local invoke -t ./cdk.out/TODOStack.template.json MyFunction
```

You can find out more about this method of local testing by following this URL: `https://docs.aws.amazon.com/serverless-application-model/latest/developerguide/serverless-cdk-testing.html`.

Ultimately, while the tooling for local serverless development might not be mature, many efforts are being made right now to close the gaps. We recommend keeping an eye on the latest tooling to evolve the way you code serverless infrastructure as the space becomes more mature.

Summary

In this chapter, we discussed how to streamline serverless development by using a local express server and LocalStack for a local cloud environment. We provided step-by-step instructions for installing and running LocalStack and `cdklocal`, configuring the DynamoDB endpoint to point to LocalStack, and importing the same function used in the deployed Lambda function to the local server. Additionally, we showed how to run commands and make requests to the local development server to test the code's functionality. The overall objective of this process is to enable the local testing and development of AWS cloud infrastructure without incurring costs or impacting production resources. In the next chapter, we will introduce the concept of **indestructible serverless application architecture (ISAA)**, a design pattern for building serverless applications that are highly resilient and can scale infinitely.

Part 4: Advanced Architectural Concepts

In this part, we will introduce the concept of **Indestructible Serverless Application Architecture (ISAA)**, and cover how **Cloud Development Kit (CDK)** and other cloud technologies enable this shift to a more reliable and maintainable future for cloud applications. We will also look at the current and future landscape with regard to CDK-like tools and see some of the open source alternatives. This part includes the following chapters:

- *Chapter 9, Indestructible Serverless Application Architecture (ISAA)*
- *Chapter 10, The Current CDK Landscape and Outlook*

Indestructible Serverless Application Architecture (ISAA)

We have uploaded modified WordPress PHP files via FTP with drag and drop. I've (Mark) worked with Vagrant, Chef, and Puppet (which was the first time I was introduced to infrastructure automation) and, later, with tools such as Docker, Terraform, Kubernetes, and Helm. As well as making the life of a DevOps engineer easier, they also allow for more complex infrastructure use cases to be addressed. AWS CDK is the next step in this evolution.

Up until this chapter, we've gone through the practicalities of using AWS CDK. For our team at Westpoint, CDK and similar tools, such as **Pulumi**, **CDKTF**, and **cdk8s** (which we will explain in further detail in the next chapter), have an important significance. This is the first time we've seen cloud infrastructure automation be Turing complete, and with the likes of Node.js, TypeScript, and the CDK standard library, we now have advanced tooling to instruct this Turing-complete machine.

If you're not familiar with the concept of Turing completeness of a programming language, in layperson's terms, a Turing-complete language can implement every potential algorithm, such as sorts, comparisons, and loops. Examples of such programming languages are C, Java, Clojure, and TypeScript. In contrast, non-Turing-complete languages are ones that can't have some algorithms implemented in them. Examples of these are JSON, YAML, and HCL, which are extensively used in infrastructure automation tooling. CDK changes this and it has inspired others to create tooling like CDK to tackle infrastructure challenges.

New tooling allows for new ways to architect applications to achieve higher degrees of resilience, scalability, and long-term code stability. In this chapter, we will go through the concept of what our team calls **ISAA**.

In this chapter, we will learn about the following topics:

- What is ISAA?
- Principles of ISAA
- Example scenarios where ISAA can be used

Technical requirements

This chapter takes a different approach from the previous ones, as we won't be delving into any specific code. Instead, we will be providing a maintained version of the TODO app that reflects our latest understanding of the topics covered in this chapter. You can access it by following this link: `https://github.com/PacktPublishing/AWS-CDK-in-Practice/tree/main/chapter-9-indestructible-serverless-application-architecture-isaa`.

The Code in Action video for this chapter can be viewed at: `https://packt.link/7k4G8`.

What is ISAA?

The rate of innovation for new tooling either for or inspired by CDK is so mind-bendingly high that nearly every day, we come across a new CDK construct, library, or concept that deals with highly challenging infrastructure problems.

This is no surprise. New tooling creates new ways of thinking about building applications. The way React applications are configured now is different from the way Backbone.js applications were developed, organized, and architected. One could not dream of reaching the same level of resilience for microservices without Kubernetes as an orchestrator of services becoming available.

Much like microservice architecture, AWS CDK allows for a new generation of applications that have even higher degrees of resilience, scalability, and maintainability. It's a method we use at Westpoint. In fact, we oversee supporting applications that we developed two years ago that need very little support, limited to bug fixes and data clean-ups at times. As long as AWS as an entity exists, these applications will continue working without a hitch. We call this type of cloud software architecture the **Indestructible Serverless Application Architecture (ISAA)**.

Before going further, please consider that each of the topics discussed here could well be a book of its own. Our aim with this chapter is to show you the way our team – with our limited knowledge, of course – is starting to think about building robust web applications. We're pretty sure there are further discussion points and downsides to the claimed upsides, but for us, the results so far speak for themselves.

A deployed software system left to its own devices will get increasingly unstable. The reasons can vary. Here are some example scenarios:

- The server where you deployed your web application fails because it runs out of disk space as a result of application log accumulation
- A library module used by your application has a memory leak, which will demand an increasing amount of system memory and cause a crash
- Your web app is inaccessible over TLS because the TLS certificate for your domain expired
- You own an analytics application and the daily ETL service chokes on data since the data from the previous day took more than 24 hours to be processed

There are a ton of other curveballs the universe throws at your code. In fact, the measure of chaos in software – or **software entropy** – is closely related to the concept of entropy in thermodynamics (forgive me for bringing physics into this), which states the following:

"Every system left to its own devices always tends to move from order to disorder."

The key phrase here is *left to its own devices*. With ISAA, we are refusing to leave the system (or in our case, the software) to its own devices. We will use the power of AWS's serverless workhorse to shape our application in such a way that our applications become as reliable as the AWS services powering them. AWS charges more for their serverless services in comparison to their managed services, but the charge incurred by extra developer time to write up procedures that provide the given level of resilience would be a lot higher. AWS is able to provide the same service for a lower cost because it can apply the economy of scale to it. Using ISAA, we are, in essence, subcontracting DevOps to AWS.

Let's look back at the examples that would increase the instability of our software deployment and see how ISAA would deal with them:

- Don't accumulate logs on the server. Send them to a logging service such as CloudWatch. Set the log expiry date to a month or whatever frequency you'd like to go back and review logs so that you won't be indefinitely charged for storing logs.

- Base the application on AWS Lambda. Every feature of the application will call a Lambda function that will serve the request and get immediately destroyed. The next Lambda function will be spun up with fresh resources. Basically, every call becomes isolated.

- Issue certificates with the AWS Certificate Manager service and set renewal alarms in your CDK code or configure automated certificate renewal.

- Use AWS Glue, which allows virtually unlimited scaling for your ETL jobs. Get the jobs done quickly. Or even better, base the analytics on Kinesis Data Analytics, which is a streaming serverless map/reduce platform that provides real-time results.

ISAA has only been practically possible with AWS CDK.

Consider the second example. The solution would perhaps require 100 Lambda functions to be provisioned. If we had to define each in the Terraform or CloudFormation way, it would result in unmaintainable amounts of configuration files. All the other cases mentioned would also be heavy on the configuration and difficult to follow because of the lack of followable logic in the likes of HCL, YAML, or JSON.

So, how does ISAA differ from a serverless architecture? Well, it's a subset of serverless architecture but with some additional principles baked in, which we will be covering next.

ISAA principles

So how does one make an application virtually indestructible? Let's review some of the principles that ensure that ISAA is indestructible.

A fuller stack

With the advent of AWS CDK, the worlds of full stack engineers and DevOps engineers have begun to merge into one. CDK allows developers to code their infrastructure, giving them more options to build resilient applications. This means that developers now have the ability to take responsibility for their infrastructure code, rather than simply piling things up in a container and passing it on to a DevOps team to deploy.

Gone are the days when developers were only responsible for writing application code and DevOps teams were responsible for deploying and managing the infrastructure. With CDK, developers can now write both application code and infrastructure code and have more control over the entire application life cycle. This allows for a more streamlined and efficient development process, as well as a greater level of ownership and accountability for the final product.

As a result of this shift, the distinction between DevOps and developers is becoming less clear. ISAA requires this switch in the mentality of dev teams. Every developer should be able to work on the *fuller stack* of the application. There would obviously be more DevOps engineers in teams, but everyone needs to at least be aware of how the application ties together. So, of course, not every developer is expected to be able to master AWS, but if you know AWS and CDK alongside other coding skills, you will be building bulletproof cloud applications.

Serverless

By nature, ISAA is serverless. So far in this book, we've complained enough about maintaining complex infrastructure. Say what you will but we don't like doing it and we know many more like-minded people are out there. A world free of this burden is only achievable via serverless.

Our problem with the state of serverless right now is the lack of openness. We would ideally be free to choose which hardware to run our services on. Every serverless implementation is turning into a proprietary platform. Yes, Richard Stallman warned us of this. This is indeed a worry. Our argument for using serverless is that it is driving innovation in open alternatives as well. We don't think we can create openness without poking at proprietary implementations first. We will see some of these alternatives in *Chapter 10*.

Simplicity

An ISAA application should be simple to define and maintain. Developers should easily be able to add features to the application and deploy them with ease, not worrying about complex wiring and data migrations. This means fewer moving parts are desired. An example of this is the DynamoDB single-table design, which simplifies the database design and improves scalability and performance, often leading to applications just needing one database.

> **Single-table design**
>
> Single-table design is a way of organizing data in a NoSQL database that makes it easy to scale and manage. Instead of having multiple tables for different types of data, you just keep it all in one table. Each item in the table has a unique ID and another key to sort and filter the data. This makes it easy to find what you're looking for. While there are downsides to having data repeated in multiple places to make it faster to access with a single-table design, dealing with these downsides becomes a lot easier with AWS CDK.
>
> Alex DeBrie has a few excellent articles explaining the single-table design at `https://www.alexdebrie.com/posts/dynamodb-single-table/`.

Another example would be the use of AWS Cognito for **identity and access management** (**IAM**). We've seen IAM products that have taken months to be deployed and are highly difficult to integrate with. AWS Cognito supports the most popular protocols such as **OAuth 2.0** and **OpenID Connect**, so you can use the power of stateless tokens (JWTs) without having to deal with the hassle of running a resilient IAM service in your stack. In addition, Cognito can serve as an authentication mechanism for several AWS services, including REST and GraphQL APIs on API Gateway.

Event-driven architecture

A soft prerequisite of ISAA is **event-driven architecture** (**EDA**). EDA is a way of designing software where the main focus is on events happening and how to respond to them. The events can be anything from small data changes to big business processes and DevOps-related triggers.

To understand this point better, let's consider the following feature request for an e-commerce application.

Create an API endpoint that does the following:

- Accepts the JSON payload of an order
- Saves the order in the database
- Sends a notification to the user

The following diagram explains the traditional way of dealing with this sequence of actions:

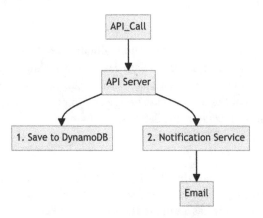

Figure 9.1 – Traditional way of dealing with orders and notifications

With ISAA, we would approach this differently. We use a feature of DynamoDB named **DynamoDB Streams** and Lambda triggers. DynamoDB Streams is a powerful feature that allows you to easily trigger a Lambda function when a certain event occurs in your DynamoDB table. This means that whenever an item is added, updated, or removed from your table, a stream event will be generated and sent to a Lambda function of your choice. This is extremely useful for creating real-time data processing pipelines, or for performing actions such as sending notifications or updating other systems in response to changes in your data. The best thing is that the triggering of the Lambda function is done automatically and seamlessly, so you don't have to worry about polling or manually triggering the function. And the most interesting thing is that you can even filter the events based on specific values in the data and choose which events trigger the Lambda function. This gives you a lot of flexibility and power to handle your data in real time, and it's the event-driven way of doing things.

So going back to the API flow, here is the event-driven way of doing things:

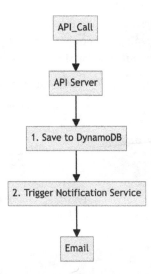

Figure 9.2 – The event-driven way of sending notifications for orders

So, in this sequence, we have decoupled the notification mechanism from the service action saving the order. In this instance, we are triggering a notification service that would be a Lambda function based on changes in a certain type of record, which, in this case, is the addition of an order. And obviously, using CDK, this is a breeze:

```
import * as cdk from 'aws-cdk-lib';
import * as dynamodb from '@aws-cdk/aws-dynamodb';
import * as lambda from '@aws-cdk/aws-lambda';
import * as cdk_dynamo_stream_lambda from '@aws-cdk/aws-dynamodb-
streams-lambda';

export class SingleTableStack extends cdk.Stack {
  constructor(scope: cdk.App, id: string, props?: cdk.StackProps) {
    super(scope, id, props);

    // Create the DynamoDB table
    const table = new dynamodb.Table(this, 'single-table-db', {
      partitionKey: { name: 'pk', type: dynamodb.AttributeType.STRING
},
      stream: dynamodb.StreamViewType.NEW_IMAGE,
    });

    // Create the Lambda function
    const handler = new lambda.NodejsFunction(this, 'Handler', {
      runtime: lambda.Runtime.NODEJS_14_X,
      entry: lambda.Code.fromAsset('lambda-src'),
      handler: 'index.handler',
    });

    // Create the DynamoDB stream event source and make it trigger
only when an order item gets added
    new cdk_dynamo_stream_lambda.DynamoEventSource(table, {
      startingPosition: lambda.StartingPosition.TRIM_HORIZON,
      batchSize: 1,
      enabled: true,
      onEvent: handler,
      filter: {
        type: "prefix",
        prefix: "order/",
      },
    });
  }
}
```

Neither of us is anywhere near the best programmer out there; we're sure the smarter bunch of our industry will come up with ingenious patterns using what is now possible with CDK, but given what we wanted to achieve, the preceding code is ridiculously simple. We will also mention that DynamoDB Streams supports exactly once message delivery guarantees. If you've worked with the concept of messaging and queuing systems, you will know how insanely difficult this is to achieve on a highly reliable basis. Have you ever had to maintain a Kafka cluster? If so, you will have covered at least 30% of the total current usage of Kafka without even realizing it.

This decoupling is handled on the infrastructure level. Now, with a Turing-complete infrastructure engine, this also becomes part of the job of the developers. Trust me, this is good for our profession.

Challenges with ISAA

Given the serverless nature of ISAA, whatever challenges serverless currently has and will be found to have in the future are automatically inherited by ISAA.

Runaway costs

One of the challenges with building indestructible serverless applications is the potential for runaway costs. Because serverless architectures rely on paying for usage, it can be easy to accidentally spin up a large number of resources or have a function that runs too frequently, leading to unexpected and potentially large bills.

However, there are several ways to mitigate this risk and keep costs under control. One approach is to use monitoring and analytics tools to track usage and identify areas where costs can be reduced. For example, you can use services such as **AWS CloudWatch** or **AWS Cost Explorer** to track the number of invocations and the duration of your functions and use this information to optimize their performance. Additionally, you can set up alerts to notify you when your usage or costs reach a certain threshold.

You also can make use of the native service, **AWS Budgets**, which can help you to set cost limits; it will notify you when you reach a certain threshold so you can take action before things get out of hand.

Overall, while the potential for runaway costs is a real concern with ISAA, it is manageable with careful planning, monitoring, and optimization. Of course, all the mitigations we mentioned can also be executed using AWS CDK. Yes, you can code these mitigations, budget alarms, and alerts into your application stack.

In fact, you can create your own optimization constructs and use them across all your projects. The possibilities are endless. Here is an example of such a construct:

```
https://github.com/cdk-patterns/serverless/tree/main/the-lambda-
power-tuner
```

Lack of openness

We've dedicated the entire next chapter to this, but we think it's pretty obvious that with the current way of implementing ISAA on AWS, you are pretty much locked into a vendor with AWS.

We often get comments from more open software-oriented colleagues about the lack of openness with CDK and betting everything on AWS. We understand their concerns, but our response to them, in short, would be *"Read Chapter 10."* A slightly longer answer is that while open technologies for serverless are being developed, still, in a lot of cases, going with serverless platforms such as AWS makes a lot of business sense.

Advantages of ISAA

We don't necessarily split atoms in our organization, but we develop extremely reliable business applications. Businesses are seeing the huge advantages of serverless; combine that with event-driven architecture and a few other tweaks, and the result is that we are right now running numerous serverless applications. We find that sometimes we leave code that perfectly suits a problem running for so long without issues that we forget how it worked in the first place – essentially, realizing that we have to document our software much better going forward. But hey, we'll take that over sleepless nights trying to resolve scaling issues any day.

Our resources, the most important of which is our time, are not then spent on keeping software running but on shipping new features and working on new projects and ideas. Our engineers are a lot happier, more engaged, and creative and our customers love the reliability of the systems we produce.

Another aspect of ISAA is cost efficiency. We know we raised runaway costs as a potential negative point but given the serverless nature of this kind of architecture, and if planned right, you could run applications in a very cost-efficient way. If you have an API that serves a number of requests in the thousands or even millions, having a serverless backend can potentially be a lot more cost-efficient than running virtualized servers.

And finally, scale. If implemented right, your application stack will be able to handle large numbers of requests with a high degree of variance with ease. You have the entire power of AWS's ecosystem at your disposal to serve your customers at any time of the day in multiple regions and various availability zones.

Summary

In this chapter, we learned about ISAA and how it could help you achieve high levels of software resilience and scaling. Our hope has been to share with you the latest way we use CDK at our organization to deliver higher-quality software. In the next chapter, we look at the general landscape of IaC with CDK and other, CDK-like projects that could help us reach a degree of openness while, at the same time, still producing bulletproof apps.

10

The Current CDK Landscape and Outlook

So far in this book, we have learned how AWS CDK has revolutionized the world of **infrastructure as code (IaC)**. It has enabled developers to express cloud infrastructure using familiar programming languages such as Python, JavaScript, and TypeScript. It has redefined how we think about IaC. Developers can now programmatically define and manage infrastructure in a more flexible, scalable, and automated manner. This has enabled new use cases and allowed organizations to adopt DevOps practices more quickly and efficiently.

CDK has also allowed for new and advanced ways of spinning up infrastructure, giving organizations the ability to provision and manage their cloud resources in a more efficient and streamlined manner. As we proceed through this chapter, we will delve into some of the specific patterns that CDK has enabled and discuss how they can be leveraged to simplify and optimize the process of managing cloud infrastructure.

CDK has not only streamlined the provisioning of AWS services. It has also opened up a new realm of thought and approach to interacting with multiple other cloud providers and infrastructure tools. This has resulted in the creation of multiple offshoot projects, which, while they may not be within the scope of this book, are worth mentioning, as they can be used in conjunction with CDK in your projects.

Here's what we will learn about in this chapter:

- Some of the advanced ways CDK is put to work

- Alternatives to AWS CDK and the open source landscape

- How ISAA could be achieved using the current open source platforms and tooling

Advanced ways that CDK is leveraged

In the last chapter, we covered ISAA, our preferred ways of creating cloud applications that stand the test of time so well that we seldom need to conduct any infrastructure maintenance procedures. As CDK adoption is increasing, we are seeing interesting infrastructure patterns emerge. In this section, we will cover some of these new and interesting ways companies are using CDK to respond to infrastructure challenges that previously required heavy development work that made them out of reach for start-ups and other small tech businesses.

Creating multiple CDK stacks

For the sake of simplicity, we have so far been mainly working with a single CDK stack in each CDK app. As mentioned before, AWS CloudFormation has limit of 500 resources per stack. You might think this is a large number, but when you deal with serverless applications and deploy many Lambda functions and their dependencies such as roles and layers, you will hit this limit.

One way around this limit is to split the stack into multiple stacks. You would have to be strategic with how you split the resources. This has to be done in a way that minimizes intra-stack dependencies since such dependencies are managed by exporting certain values as outputs for a certain stack and then importing them from another stack. Here is how this works in practice.

Imagine we have two CDK stacks named StackA and StackB. StackA generates an S3 bucket with a one-of-a-kind name, and we intend to utilize this distinct name as a parameter in StackB to generate an SNS topic that subscribes to S3 events:

```
// Define StackA
const stackA = new Stack(app, 'StackA');
const s3Bucket = new s3.Bucket(stackA, 'MyBucket', {
  bucketName: `${app.node.tryGetContext('uniqueBucketName')}-my-
bucket`,
});

// Define StackB
const stackB = new Stack(app, 'StackB');
const bucketName = stackA.getOutput('BucketName');

new sns.Topic(stackB, 'MyTopic', {
  displayName: 'My SNS Topic',
  subscription: [{
    protocol: sns.SubscriptionProtocol.EMAIL,
    endpoint: 'example@example.com',
  }],
  topicName: 'my-topic',
  filterPolicy: {
    bucketName: {
```

```
      'Fn:Equals': [
        { 'Fn:ImportValue': `${bucketName}-MyBucketName` },
        s3Bucket.bucketName,
      ],
    },
  },
});
```

In this instance, we specify `StackA` to create a distinct Amazon S3 bucket using the `uniqueBucketName` parameter from the CDK context. Afterward, we define `StackB` to generate an Amazon SNS topic that receives S3 events. To obtain the exclusive bucket name from `StackA`, we use the `getOutput` method to retrieve the name as an output from `StackA`. Eventually, we implement the bucket name as a filter policy for the SNS topic's subscription to exclusively get S3 events from that specific S3 bucket.

This is OK for when you have only one dependency between your stacks, but it becomes complicated and quickly falls apart if the number of dependencies is high. Here are some general guidelines to follow when splitting CDK resources into multiple stacks:

- **Group resources by lifecycle**: Group resources that have a similar life cycle, such as those that are created, updated, or deleted together, into the same stack.

- **Minimize cross-stack dependencies**: Minimize the dependencies between stacks to reduce the risk of errors and simplify the deployment process.

- **Reuse resources where possible**: Identify resources that can be reused across multiple stacks, such as network infrastructure or security groups. This can help to reduce duplication of effort and simplify the management of the application's infrastructure.

- **Consider security and compliance requirements**: Ensure that each stack is deployed in a secure and compliant manner and that resources are not shared across stacks unless necessary.

- **Group stateful resources in the same stack**: Separating stateful resources such as databases from stateless resources allows for the stateless stack to be easily destroyed or created without risking data loss. It's worth noting that stateful resources are more sensitive to being replaced, so it's not recommended to nest them inside constructs that may be renamed or moved around. When creating multiple stacks and separating resources, this is important to keep in mind.

Dynamic provisioning

Dynamic provisioning is a powerful pattern made possible by AWS CDK. It allows developers to create an infrastructure that can be automatically adjusted based on changing conditions or requirements. This means that as your application grows and evolves, your infrastructure can grow and change right along with it.

Dynamic provisioning is all about being able to respond quickly to changes in demand. For example, if your application suddenly experiences a spike in traffic, dynamic provisioning can automatically spin up additional resources to handle the increased load. This can help to ensure that your users have a smooth and uninterrupted experience, even during times of high demand.

At its core, dynamic provisioning is about maximizing efficiency and reducing waste. Instead of having to manually provision new resources every time your application grows, you can automate the process and have the right resources available exactly when you need them. This can not only save you time and effort, but it can also help you to avoid overprovisioning, which can result in unnecessary costs. So, if you're looking for a way to stay ahead of the curve and stay on top of your infrastructure needs, dynamic provisioning is definitely a pattern you should consider.

The technical details of this pattern are very interesting. Here is an AWS blog article covering this solution in depth – `https://aws.amazon.com/blogs/developer/how-vendia-leverages-the-aws-cdk-to-dynamically-provision-cloud-infrastructure/`.

CDK for larger organizations

When starting to use AWS CDK, care must be given when planning how the CDK projects, general cloud architecture, and infrastructure provisioning is to be done. For smaller companies, it might be just one or two people, while larger companies might have a full **Cloud Center of Excellence** (**CCoE**). A CCoE is a team of experts within an organization that helps make sure the company is using cloud computing in the best way possible. The team helps create and implement best practices for cloud usage, makes sure everyone in the organization is using the cloud in a way that makes sense for the business, and ensures that the company's cloud strategy aligns with its goals. The ultimate goal of a CCoE is to help the company save time and money while taking advantage of the benefits of cloud computing.

An AWS **landing zone** is an important aspect of the operations of the CCoE. A landing zone is a preconfigured AWS environment that helps you set up a secure and scalable infrastructure quickly. It's like a starter kit for creating new AWS accounts or migrating existing workloads to AWS. It provides a set of best practices and guidelines for creating and managing AWS accounts, making it easier to enforce policies and standards across the entire AWS infrastructure. To manage your multi-account system, you can use **AWS Control Tower**, which provides a single user interface to configure and manage your entire system. Developers can use their own accounts for testing and deployment, and for each isolated account, the CDK applications can be deployed to various environments (dev, stg, prod, etc.).

And yes, you guessed it, we can define these landing zones and configure AWS Control Tower using CDK. This goes back to the monorepo idea we introduced in *Chapter 2*. The concept of monorepo could extend to the organizational level in such a way that you can define the AWS account-level policies, restrictions, and guidelines using landing zones configured with CDK. Various projects could share the same CDK constructs, and various stacks could import outputs of other stacks.

CDK for other platforms

In *Chapter 9*, when we covered the advantages and disadvantages of ISAA, we promised that we would cover some CDK-like libraries out there that are on the path to promoting openness in the serverless field. Let's first cover some CDK alternatives.

Pulumi

Pulumi (`https://www.pulumi.com/`) is the most complete CDK alternative we've seen. Its API is slightly different from CDK, but at the end of the day, it produces the same results on AWS. Like CDK, it's an Apache-licensed open source project with a service, and its cost is based on the number of L2 or L3 components Pulumi *manages* for you. But it comes with a great advantage: as well as targeting AWS, it targets **Google Cloud Platform** (**GCP**), Azure, and Kubernetes. We believe they might even add more cloud providers to the list in the future.

> **Which came first?**
>
> CDK was released in 2018, earlier than Pulumi. But Pulumi as a company was established a year earlier than CDK's release, so perhaps they also independently had the idea of provisioning cloud infrastructure with code.

Our team has used Pulumi for some side projects to experiment with the technology, and the feedback has generally been positive other than getting used to some of the quirks that come with Pulumi's toolset. If being cloud agnostic is a requirement for your organization, we definitely recommend giving Pulumi a try.

> **Cloud agnostic**
>
> Being cloud agnostic means that an application or system is designed to work across multiple cloud computing platforms or environments. It allows for greater flexibility and avoids vendor lock-in. How much the likes of Pulumi help achieve this is, however, questionable. In our opinion, the only tool on the market that delivers cloud agnosticism right now is Kubernetes. More on this will be covered in the *ISAA on Kubernetes* section.

Here is a simple Pulumi app that creates an ECS cluster and serves an `nginx`-based service just like the one we deployed in *Chapter 1*:

```
import * as pulumi from "@pulumi/pulumi";
import * as aws from "@pulumi/aws";

// Create an Amazon ECS cluster
const cluster = new aws.ecs.Cluster("my-cluster");

// Create a task definition for the ECS service
```

```
const taskDefinition = new aws.ecs.TaskDefinition("my-task-
definition", {
    containerDefinitions: [{
        name: "my-container",
        image: "nginx",
        portMappings: [{
            containerPort: 80,
            hostPort: 80,
        }],
    }],
});

// Create an Amazon ECS service with the task definition
const service = new aws.ecs.Service("my-service", {
    cluster: cluster.id,
    desiredCount: 1,
    taskDefinition: taskDefinition.arn,
});

// Export the URL of the ECS service
export const serviceUrl = pulumi.interpolate`http://${service.
loadBalancers[0].dnsName}`;
```

You can see that given your newly acquired CDK knowledge, understanding and writing Pulumi code can be relatively easy.

Additionally, Pulumi has its own vast library of constructs and providers, which can be accessed via the Pulumi registry:

```
https://www.pulumi.com/registry/
```

In our opinion, Pulumi's registry is superior to Construct Hub's. For the poor souls out there that have had to spin up and manage Kafka clusters, we have this Pulumi code snippet for you:

```
import * as kafka from "@pulumi/kafka";

const topic = new kafka.Topic("topic", {
  name: "sample-topic",
  replicationFactor: 1,
  partitions: 4,
});
```

Nice huh? Pulumi can also target Kubernetes. Follow this link to read more – `https://www.pulumi.com/docs/get-started/kubernetes/`.

CDKTF

As the name suggests, **CDKTF** (`https://developer.hashicorp.com/terraform/cdktf`) combines the concept of CDK with the Terraform platform. Since Terraform already has an engine that tracks the state of a cloud application, CDKTF uses the same engine but adds a programming layer just like CDK does to CloudFormation.

The GitHub repository (`https://github.com/hashicorp/terraform-cdk`) shows lots of signs of activity, so we're assuming HashiCorp is taking the matter seriously. For teams that are already on Terraform, moving to CDKTF makes great sense because ultimately, it generates Terraform HCL files as output, which works with other Terraform stacks. Just like Pulumi, the CDKTF API is different from CDK, but it's all well documented, and you can use the same concepts you've learned so far in this book and apply them to your projects using CDKTF. Here is a simple example of how a CDKTF stack is written:

```
import { Construct } from 'constructs';
import { App, TerraformStack } from 'cdktf';
import { AwsProvider, S3Bucket } from './.gen/providers/aws';

class MyStack extends TerraformStack {
  constructor(scope: Construct, name: string) {
    super(scope, name);

    new AwsProvider(this, 'aws', {
      region: 'us-west-2',
    });

    new S3Bucket(this, 'my-bucket', {
      bucket: 'my-bucket',
    });
  }
}

const app = new App();
new MyStack(app, 'my-stack');
app.synth();
```

As you can see, the API is practically the same. The main difference is that given the cloud agnosticism of Terraform, constructs are organized into modules for different providers, one of which is AWS. Here is the exact same stack on GCP:

```
import { Construct } from 'constructs';
import { App, TerraformStack } from 'cdktf';
import { GoogleProvider, StorageBucket } from './.gen/providers/
google';
```

```
class MyStack extends TerraformStack {
  constructor(scope: Construct, name: string) {
    super(scope, name);

    new GoogleProvider(this, 'google', {
      project: 'my-gcp-project-id',
      region: 'us-west1',
    });

    new StorageBucket(this, 'my-bucket', {
      name: 'my-gcs-bucket',
    });
  }
}

const app = new App();
new MyStack(app, 'my-stack');
app.synth();
```

Just like Pulumi, Terraform's cloud agnosticism makes CDKTF very powerful. This means you can have resources on multiple cloud providers within the same app. Say you run a Kubernetes cluster on GCP but want to also use DynamoDB on AWS. With CDKTF and Pulumi, you would be able to easily do this.

CDK8S

Our favorite CDK-inspired project right now is CDK8S (`https://cdk8s.io/`). Made by AWS, it's an open source CDK tool implemented on top of Kubernetes, an open source platform. The API library works in very similar ways to AWS CDK, although it provisions Kubernetes services. Here is a CDK8S code that creates an `nginx`-served web app that is replaced across three pods:

```
import { Construct } from 'constructs';
import { App, Chart } from 'cdk8s';
import { Deployment } from './imports/k8s';

class MyChart extends Chart {
  constructor(scope: Construct, name: string) {
    super(scope, name);

    const deployment = new Deployment(this, 'my-deployment', {
      spec: {
        replicas: 3,
        selector: {
          matchLabels: {
            app: 'my-app',
```

```
          },
        },
      template: {
        metadata: {
          labels: {
            app: 'my-app',
          },
        },
        spec: {
          containers: [
            {
              name: 'my-container',
              image: 'nginx',
              ports: [{ containerPort: 80 }],
            },
          ],
        },
      },
    },
  });
  }
}


const app = new App();
new MyChart(app, 'my-chart');

app.synth();
```

A CDK8S chart is much like a Helm chart, if you are familiar with the concept. This in turn translates to a stack or L3 construct in the CDK universe. In fact, since Helm has been around for a while, and a lot of services have Helm charts that spin them up and manage them, CDK8S supports the importation of Helm charts into your code. So, again, like CDKTF, if you're part of the Kubernetes world, you can apply what you've learned in this book to Kubernetes using CDK8S.

The following code does the same thing as the previous one but uses a Helm chart to define nginx instead:

```
import { Chart, Helm } from 'cdk8s';
import { Construct } from 'constructs';


export class MyChart extends Chart {
  constructor(scope: Construct, name: string) {
    super(scope, name);

    // Import the nginx Helm chart
    const nginx = Helm('nginx', {
```

```
    repository: 'https://charts.bitnami.com/bitnami',
    chart: 'nginx',
    version: '8.1.1',
  });

  // Set the number of replicas for the nginx deployment
  nginx.values.spec.replicas = 3;
  }
}
```

Kubernetes veterans will know that this is an extremely simple way of achieving the intended objective. You can even combine CDK8S with CDK. Have your AWS-level resources, including your EKS cluster, defined in CDK, and the Kubernetes services defined in CDKTF. In your CI/CD process, compile and synthesize the CDK8S project and use the output in your CDK EKS definitions.

Serverless Stack Toolkit

Serverless Stack Toolkit (SST) is a framework for building serverless applications on AWS, much like Pulumi and CDKTF. It provides a simplified developer experience with abstractions and easy-to-use APIs, while still allowing for customization and flexibility. SST allows developers to quickly build serverless applications using familiar programming languages such as JavaScript or TypeScript. It also includes helpful features such as local development, automatic IAM permissions, and integrated deployments.

SST can be used as a substitute for CDK, as it is easier to use, offers a higher level of abstraction, and still allows for customization and flexibility. While CDK provides a powerful way to define IaC using familiar programming languages, it can be complex and has a steep learning curve. SST offers a simpler and more streamlined approach to building serverless applications on AWS, which can be beneficial for smaller projects or for developers who are new to AWS and serverless development. In comparison to the other alternatives, SST focuses on serverless-first development on AWS, while Pulumi offers a more general-purpose approach that supports multiple cloud providers. CDKTF, meanwhile, provides a bridge between AWS CDK and Terraform.

Here's an example of how to deploy a React app from SST's repository:

```
export function ExampleStack({ stack }: StackContext) {// Deploy our
React app
const site = new StaticSite(stack, "ReactSite", {
  path: "packages/frontend",
  buildCommand: "npm run build",
  buildOutput: "build",
  environment: {
    REACT_APP_API_URL: api.url,
```

```
  },
});

// Show the URLs in the output
stack.addOutputs({
  SiteUrl: site.url,
  ApiEndpoint: api.url,
});
}
```

ISAA on Kubernetes

Let's go back to the simple, serverless, ISAA-modeled TODO application that we created in *Chapter 8*. We'll summarize some of the key components that made it indestructible:

- **AWS CDK**: We used CDK to programmatically define our infrastructure and easily spin up otherwise complicated stacks

- **CI/CD**: The application had a robust CI/CD pipeline built on CodeBuild and CodePipeline

- **API Gateway**: There was an API layer that dealt with ingress/egress and access

- **Lambda functions**: These were serverless functions that we ran with the addition of `StepFunction` state machines

- **DynamoDB** and **DynamoDB Streams**: We used the power of DynamoDB, DynamoDB Streams, and single-table design to achieve an event-based architecture

We believe we are close to the point of being able to achieve ISAA in a cloud-agnostic and open source way. Let's review the same points and see what open source alternatives there are for each:

- **AWS CDK**: We have the likes of CDK8S, Pulumi, SST, and CDKTF, which at times are superior to AWS CDK.

- **CI/CD**: We have Jenkins X, Tekton, and Argo CD. Again, in certain aspects, these are superior to AWS tooling.

- **API Gateway**: We have the likes of Nginx, Ambassador, and many more, but we've generally found those options lacking the completeness of AWS's API Gateway and its interoperability with the likes of IAM, Lambda, and AWS Cognito. Kubernetes Gateway API is a new arrival, and it has a lot of potential: `https://gateway-api.sigs.k8s.io`.

- **Lambda functions**: A lot of exciting new developments have happened in this space. Projects such as Fission (`https://fission.io`) and OpenFaaS (`https://www.openfaas.com`) are production-ready alternatives to Lambda. Step functions, however, are still out of reach.

- **DynamoDB and DynamoDB Streams**: ScyllaDB is a great option in this space. DynamoDB Streams' level of integration with other serverless functions is just not there yet.

  One solution could be achieved by reading the CDC streams from ScyllaDB:

  `https://docs.scylladb.com/stable/using-scylla/cdc/cdc-streams.html`

  We could then publish them into a queue, which in turn triggers OpenFaaS functions:

  `https://docs.openfaas.com/reference/triggers/`

  This might sound daunting, but given how powerful the CDK alternatives are, this can be done relatively easily.

We are not too far off. The main missing piece is the integration between various components. AWS provides a common interface and model to deal with all these various elements and easily allows cross-service integration. We believe it's better to get things done as quickly and as efficiently as possible, hence our organization relies heavily on AWS and CDK to deliver our projects. In some instances, certain solutions have to be cloud agnostic, which can happen for multiple reasons outside the scope of this book. One such case for us was that we had to be able to provide the same cloud solution on-premises. In that scenario, while making use of serverless Fission functions on Kubernetes, we kept most of the stack in containers and pods. Going forward, our aim will remain to build a comprehensive and serverless solution on open platforms.

Where to go from here

Using AWS CDK or other CDK alternatives effectively requires a deeper insight into architectural practices, AWS services, platforms such as Kubernetes, and the latest developments in the field. The coding examples and the architectural practices used in this book are just a starting point. There are teams hard at work building incredible things using the technology. Here are some starting points for you to expand your knowledge and remain up to date:

- We highly recommend you get at least one AWS certification, be it foundational, associate, or a higher level: `https://aws.amazon.com/certification/`

  This will help you immensely with getting to know AWS's services and cloud best practices.

- The AWS Well-Architected Framework is also an incredibly useful resource to learn about the fundamentals of a great cloud application: `https://aws.amazon.com/architecture/well-architected`

  There is a serverless-focused article here: `https://aws.amazon.com/blogs/compute/building-well-architected-serverless-applications-introduction/`

- This best practices guide for CDK is full of great points to consider when developing CDK applications: `https://docs.aws.amazon.com/cdk/v2/guide/best-practices.html`

- Keep an eye on Pulumi's blog, because the team is truly on fire: `https://www.pulumi.com/blog/`

- The Kubernetes blog has the latest developments regarding the platform: `https://kubernetes.io/blog/`

- CDK Workshop is a well-designed step-by-step tutorial for AWS CDK: `https://cdkworkshop.com/`

Summary

In this chapter, we covered some of the more sophisticated usages of AWS CDK, looked at open source tooling, and briefly covered the potential of using CDK alternatives and other open source solutions to build highly resilient serverless applications.

This chapter concludes our book. It has been somewhat difficult to gather information about such a new and constantly changing technology. We hope this book has been helpful in your endeavor to build great cloud applications. We can't wait to see how AWS CDK changes the way you run cloud solutions in your organization and would love to hear back from you with any feedback regarding the book, the projects you utilize CDK for, or just a general chat on the topics.

Thank you from the Westpoint team, happy coding, and goodbye.

Index

packtpub.com

Subscribe to our online digital library for full access to over 7,000 books and videos, as well as industry leading tools to help you plan your personal development and advance your career. For more information, please visit our website.

Why subscribe?

- Spend less time learning and more time coding with practical eBooks and Videos from over 4,000 industry professionals

- Improve your learning with Skill Plans built especially for you

- Get a free eBook or video every month

- Fully searchable for easy access to vital information

- Copy and paste, print, and bookmark content

Did you know that Packt offers eBook versions of every book published, with PDF and ePub files available? You can upgrade to the eBook version at packtpub.com and as a print book customer, you are entitled to a discount on the eBook copy. Get in touch with us at customercare@packtpub.com for more details.

At www.packtpub.com, you can also read a collection of free technical articles, sign up for a range of free newsletters, and receive exclusive discounts and offers on Packt books and eBooks.

Other Books You May Enjoy

If you enjoyed this book, you may be interested in these other books by Packt:

AWS Cloud Computing Concepts and Tech Analogies

Ashish Prajapati, Juan Carlos Ruiz

ISBN: 978-1-80461-142-5

- Implement virtual servers in the cloud
- Identify the best cloud storage options for a specific solution
- Explore best practices for networking and databases in the cloud
- Enforce security with authentication and authorization in the cloud
- Effectively monitor applications in the cloud
- Leverage scalability and automation in the cloud
- Get the hang of decoupled and serverless architecture
- Grasp the fundamentals of containers and Blockchain in the cloud

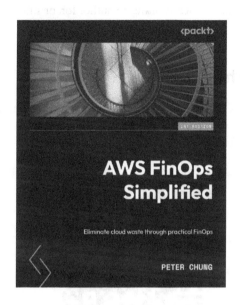

AWS FinOps Simplified

Peter Chung

ISBN: 978-1-80324-723-6

- Use AWS services to monitor and govern your cost, usage, and spend
- Implement automation to streamline cost optimization operations
- Design the best architecture that fits your workload and optimizes on data transfer
- Optimize costs by maximizing efficiency with elasticity strategies
- Implement cost optimization levers to save on compute and storage costs
- Bring value to your organization by identifying strategies to create and govern cost metrics

Packt is searching for authors like you

If you're interested in becoming an author for Packt, please visit `authors.packtpub.com` and apply today. We have worked with thousands of developers and tech professionals, just like you, to help them share their insight with the global tech community. You can make a general application, apply for a specific hot topic that we are recruiting an author for, or submit your own idea.

Share Your Thoughts

Now you've finished *AWS CDK in Practice*, we'd love to hear your thoughts! Scan the QR code below to go straight to the Amazon review page for this book and share your feedback or leave a review on the site that you purchased it from.

`https://packt.link/r/180181239X`

Your review is important to us and the tech community and will help us make sure we're delivering excellent quality content.

Download a free PDF copy of this book

Thanks for purchasing this book!

Do you like to read on the go but are unable to carry your print books everywhere? Is your eBook purchase not compatible with the device of your choice?

Don't worry, now with every Packt book you get a DRM-free PDF version of that book at no cost.

Read anywhere, any place, on any device. Search, copy, and paste code from your favorite technical books directly into your application.

The perks don't stop there, you can get exclusive access to discounts, newsletters, and great free content in your inbox daily

Follow these simple steps to get the benefits:

1. Scan the QR code or visit the link below

https://packt.link/free-ebook/978-1-80181-239-9

2. Submit your proof of purchase
3. That's it! We'll send your free PDF and other benefits to your email directly